12th International Workshop on Electric Probes in Magnetized Plasmas (IWEP 2017)

Journal of Physics: Conference Series Volume 958

Naklo, Slovenia
4 – 7 September 2017

ISBN: 978-1-5108-5849-7
ISSN: 1742-6588

Printed from e-media with permission by:

Curran Associates, Inc.
57 Morehouse Lane
Red Hook, NY 12571

Some format issues inherent in the e-media version may also appear in this print version.

Printed by Curran Associates, Inc. (2018)

For permission requests, please contact the Institute of Physics at the address below.

Institute of Physics
Dirac House, Temple Back
Bristol BS1 6BE UK

Phone: 44 1 17 929 7481
Fax: 44 1 17 920 0979

techtracking@iop.org

Additional copies of this publication are available from:

Curran Associates, Inc.
57 Morehouse Lane
Red Hook, NY 12571 USA
Phone: 845-758-0400
Fax: 845-758-2633
Email: curran@proceedings.com
Web: www.proceedings.com

Table of contents

Volume 958

12th International Workshop on Electric Probes in Magnetized Plasmas 2017 (IWEP 2017)

4–7 September 2017, Naklo, Slovenia

Accepted papers received: 16 January 2018
Published online: 12 February 2018

Preface

Papers

IOP Conf. Series: Journal of Physics: Conf. Series **958** (2018) 011001 doi:10.1088/1742-6596/958/1/011001

Foreword

International Workshops on Electric Probes in Magnetized Plasmas (IWEP) take place every second year since 1993. They are devoted to understanding and solving problems associated with electric probes in either magnetized or unmagnetized plasmas. The 12th International Workshop on Electric Probes in Magnetized Plasmas 2017 (IWEP 2017) was held in Hotel Marinšek in Naklo, Slovenia from 4th to 7th of September 2017 and was organized by the Nuclear Society of Slovenia. This Society has a long tradition of successful organization of scientific conferences and workshops. The organizers have received 18 abstracts. All the abstracts were approved by the International Scientific Committee and are published in the Book of abstracts. At the workshop 15 contributions were presented. All presentations were oral and of equal length – 30 minutes for presentation and 5-10 minutes for discussion. Eight presented contributions were submitted for publication in the proceedings. After a standard peer review procedure 7 were accepted for publication.

Committees

Members of the committees are listed in alphabetic order.

International Scientific Committee

Michael LAUX, Germany, Chair

John ALLEN, UK

Gheorghe DINESCU, Romania

Åshild FREDERIKSEN, Norway

James Paul GUNN, France

Tomaž GYERGYEK, Slovenia

Carlos HIDALGO, Spain

Mark KOEPKE, USA

Emilio MARTINES, Italy

Hans Werner MÜLLER, Germany

Tsviatko POPOV, Bulgaria

Jens Juul RASMUSSEN, Denmark

Roman SCHRITTWIESER, Austria

IOP Conf. Series: Journal of Physics: Conf. Series **958** (2018) 011001 doi:10.1088/1742-6596/958/1/011001

Carlos SILVA, Portugal

Reiner STENZEL, USA

Organizing Committee

Tomaž Gyergyek, Chair

Saša Bobič

Mateja Južnik

Jernej Kovačič

Urška Turšič

Nina Udir

Bojan Žefran

Peer review statement

All papers published in this volume of *Journal of Physics: Conference Series* have been peer reviewed through processes administered by the proceedings Editors. Reviews were conducted by expert referees to the professional and scientific standards expected of a proceedings journal published by IOP Publishing.

IOP Conf. Series: Journal of Physics: Conf. Series **958** (2018) 012001 doi:10.1088/1742-6596/958/1/012001

Current-Voltage and Floating-Potential characteristics of cylindrical emissive probes from a full-kinetic model based on the orbital motion theory

Xin Chen[1] **and Gonzalo Sánchez-Arriaga**[2]

[1 2] Bioengineering and Aerospace Engineering Department, Universidad Carlos III de Madrid, Avd. de la Universidad 30 28911, Leganés (Madrid), Spain

E-mail: [1]`xin.chen@uc3m.es`

E-mail: [2]`gonzalo.sanchez@uc3m.es`

Abstract. To model the sheath structure around an emissive probe with cylindrical geometry, the Orbital-Motion theory takes advantage of three conserved quantities (distribution function, transverse energy, and angular momentum) to transform the stationary Vlasov-Poisson system into a single integro-differential equation. For a stationary collisionless unmagnetized plasma, this equation describes self-consistently the probe characteristics. By solving such an equation numerically, parametric analyses for the current-voltage (IV) and floating-potential (FP) characteristics can be performed, which show that: (a) for strong emission, the space-charge effects increase with probe radius; (b) the probe can float at a positive potential relative to the plasma; (c) a smaller probe radius is preferred for the FP method to determine the plasma potential; (d) the work function of the emitting material and the plasma-ion properties do not influence the reliability of the floating-potential method. Analytical analysis demonstrates that the inflection point of an IV curve for non-emitting probes occurs at the plasma potential. The flat potential is not a self-consistent solution for emissive probes.

1. Introduction

Plasma potential is one of the principal parameters to be measured for understanding a wide range of plasma phenomena. For this purpose, emissive probes (EPs) are frequently used, mainly due to their simple practical implementation and robustness [1][2]. The determination of plasma potential relies on two experimental curves of EP measurements: the current-voltage (IV) characteristics and the floating-potential (FP) characteristics. When the EP is heated to emit thermionic electrons and thus to achieve certain emission level, the current-voltage characteristic is obtained by measuring the net current to the probe, for a range of different probe biases. The floating-potential characteristic is obtained by directly measuring the probe bias at the floating condition (i.e., zero net current), for different emission levels.

Theoretical difficulties for EPs arise from the cylindrical geometry and the space-charge effects that can result in a virtual cathode in front of the probe (i.e., non-monotonic potential profile). Available theoretical works commonly considered planar geometry [3][4][5]. The effects of the particle orbital motions were neglected [6], or partially developed in some asymptotic regimes and limited conditions [7][8]. Not until recently, a full-kinetic model based on the Orbital-Motion Theory (OMT) for cylindrical EPs was finally developed [9]. This work is organized as follows. In Sec. 2, the model in [9] is briefly described, with focus on the independent parameters that determine the EP characteristics. Parametric studies on the current-voltage and floating-potential characteristics are presented in Sec. 3 and Sec. 4, respectively, together with discussions on their applications. The conclusions are summarized in Sec. 5.

2. OMT for cylindrical emitters

2.1. Plasma and probe conditions

Let us consider a long emissive probe (denoted by the subscript p), with two-dimensional cylindrical geometry, radius R_p, and bias ϕ_p relative to the Maxwellian plasma at infinity that has unperturbed density N_∞, electron temperature T_e, ion temperature T_i, and ion mass m_i. As long as collisions, plasma drift, particle trapping, transient effects, and magnetic field are not significant, the OMT can be applied to model the probe-plasma interaction.

Thermionic electrons are emitted at the probe with a half-Maxwellian (HM) distribution based on the probe temperature (T_p) [10]. The thermionic current density is given by the Richardson-Dushman (RD) law as

$$J_{RD} = \lambda_{RD} A_0 {T_p}^2 \exp\left(-\frac{eW}{k_B T_p}\right) , \tag{1}$$

with W the work function in eV, T_p the probe temperature in K, λ_{RD} the material-specific correction factor, e the elementary charge, and A_0 a universal constant given by

$$A_0 \equiv \frac{4\pi e m_e k_B^2}{h^3} \approx 1.20173 \times 10^6 \mathrm{A\,m^{-2}\,K^{-2}} , \tag{2}$$

where m_e is the electron mass, k_B is the Boltzmann constant, and h is the Plank constant. Combing the HM distribution and the RD law, electrons are emitted at the probe with a density

$$\begin{aligned} N_{emp}(\lambda_{RD}, T_p, W) &= \sqrt{\frac{\pi m_e}{2 k_B T_p}} \frac{\lambda_{RD} A_0 T_p^2}{e} \exp\left(-\frac{eW}{k_B T_p}\right) \\ &= \lambda_{RD} \left(\frac{2\pi m_e k_B T_p}{h^2}\right)^{3/2} \exp\left(-\frac{eW}{k_B T_p}\right) . \end{aligned} \tag{3}$$

The dimensional parameters that determine the net current to probe are: plasma properties (T_e, T_i, m_i, and N_∞), thermionic properties of the probe (T_p, W, and λ_{RD}), and the probe geometry R_p and bias ϕ_p.

2.2. The Vlasov-Poisson system

To obtain the net current, one needs to first find out the potential distribution along the radial direction of the probe, by solving the Vlasov-Poisson system self-consistently. The dimensionless radial coordinate r is defined as the radial distance to the cylinder axis normalized by R_p. The r-dependent electric potential and particle densities are normalized as

$$\varphi(r) \equiv \frac{e\phi}{k_B T_e} , \; n_{i,e}(r) \equiv \frac{N_{i,e}}{N_\infty} , \; n_{em}(r) \equiv \frac{N_{em}}{N_{emp}} , \tag{4}$$

with the subscript $\alpha = i, e, em$ denoting for plasma ion, plasma electron, and emitted electrons, respectively. Poisson's equation for cylindrical geometry then reads

$$\frac{1}{r}\frac{\mathrm{d}}{\mathrm{d}r}\left(r\frac{\mathrm{d}\varphi}{\mathrm{d}r}\right) = -\rho_p^2 \left(n_i - n_e - \beta n_{em}\right) , \quad \varphi(r=1) = \varphi_p , \quad \varphi(r \to \infty) \to 0 , \tag{5}$$

where ions are taken as single-charged, ρ_p is the radius to Debye length ratio

$$\rho_p \equiv \frac{R_p}{\lambda_{De}} \equiv \frac{R_p}{\sqrt{\varepsilon_0 k_B T_e / e^2 N_\infty}} , \tag{6}$$

and β is the emission level

$$\beta \equiv \frac{N_{emp}}{N_\infty} = \left(\frac{\delta_p}{\delta_s}\right)^{3/2} \exp\left(-\frac{\delta_w}{\delta_p}\right) , \; \delta_w \equiv \frac{eW}{k_B T_e} , \; \delta_s \equiv \frac{h^2 N_\infty^{2/3}}{2\lambda_{RD}{}^{2/3} \pi m_e k_B T_e} , \; \delta_p \equiv \frac{T_p}{T_e} . \tag{7}$$

Based on the OMT, the densities n_α on the right-hand-side of Poisson's equation can be written out explicitly as a function of $\varphi(r)$ according to Vlasov characteristics [9] [11]. The Vlasov-Poisson system then becomes a single integro-differential equation that is to be solved for $\varphi(r)$.

The particle density n_α at any radius r can be found by integrating the local distribution function f_α over the (v_r, v_θ) velocity space,

$$n_\alpha(r) = \int\int f_\alpha (r, v_r, v_\theta) \, \mathrm{d}v_r \, \mathrm{d}v_\theta , \tag{8}$$

where the normalizations applied are $2k_B T_\alpha f_\alpha / (m_\alpha N_{\infty,emp}) \to f_\alpha$ for distribution function, $\sqrt{m_\alpha / 2k_B T_\alpha} v_r \to v_r$ for radial velocity (positive outwards), and $\sqrt{m_\alpha / 2k_B T_\alpha} v_\theta \to v_\theta$ for azimuthal velocity. The Vlasov equation conserves the distribution function along particle orbits. Due to the Maxwellian plasma faraway from the probe and the HM electrons emitted at the probe, the local distributions for the particles existing at r are

$$f_\alpha(r, v_r, v_\theta) = H_\alpha \exp\left(-{v_r}^2 - {v_\theta}^2 - u_{\alpha r}\right) / \pi , \quad H_{i,e} = 1 , \quad H_{em} = 2 , \tag{9}$$

with the normalized *electric potential energy* $u_{\alpha r}(r)$ being defined for each species as

$$u_{er}(r) \equiv -\varphi(r), \quad u_{ir}(r) \equiv \frac{\varphi(r)}{\delta_i} \equiv \frac{\varphi(r)}{T_i/T_e} , \quad u_{emr}(r) \equiv \frac{\varphi_p - \varphi(r)}{\delta_p} . \tag{10}$$

For each specie, this electric potential energy is taken relative to the *origin* of the particle (infinity for $\alpha = i, e$ and probe for $\alpha = em$).

Due to the cylindrical symmetry of the problem and its stationary character, the normalized *angular momentum* and the normalized *transverse total energy*,

$$l_\alpha \equiv r v_\theta, \quad \epsilon_\alpha \equiv v_r^2 + v_\theta^2 + u_{\alpha r} , \tag{11}$$

are also conserved along particle orbits. They not only suffice to *characterize* the orbits but also allow transforming the density integration over the $v_r v_\theta$-space to the ϵl-space. After applying change of variables $(v_r, v_\theta) \to (\epsilon_\alpha, l_\alpha)$, Eq. (8) becomes

$$n_\alpha(r) = \int\int_{2D_r} \frac{f_\alpha (r, v_r, v_\theta)}{2\sqrt{l_{\alpha r}^2(r, \epsilon_\alpha) - l_\alpha^2}} \mathrm{d}\epsilon_\alpha \, \mathrm{d}l_\alpha , \tag{12}$$

$$l_{\alpha r}^2(r, \epsilon_\alpha) \equiv {l_\alpha}^2 + r^2 v_r^2 = r^2 (\epsilon_\alpha - u_{\alpha r}) . \tag{13}$$

The domain of integration $(2D_r)$ should contain all the values of the orbital invariants $(\epsilon_\alpha, l_\alpha)$ for the particles present at r, with the factor 2 for considering only positive l_α and $l_{\alpha r}$ hereafter.

For each species, without trapped particles, the r-dependent domain D_r is composed by two sub-domains, $D_r = D_{1r} + D_{2r}$: 1) The sub-domain D_{1r} stands for the particles that are able to overcome all the barriers to arrive at r for the first time; 2) the other sub-domain D_{2r} stands for the particles that are reflected back to r and appear at that radius for the second time. For plasma species, D_{1r} and D_{2r} corresponds to *incoming* and *outgoing* particles, respectively (vice versa for emitted electrons).

According to Eq. (13), a particle is certainly *forbidden* at r if ${v_r}^2 > 0$ can not be satisfied. Consequently, for a particle to exist at r, it is *necessary* (but not *sufficient*) to have: a) an adequately large energy as $\epsilon_\alpha \geq u_{\alpha r}(r)$; b) for any energy in this range, a small enough angular momentum as $l_\alpha \leq l_{\alpha r}(r, \epsilon_\alpha)$. If the orbital characteristics (ϵ and l) of a particle can fulfil the *necessary* condition at any r' that is between r and the *origin* of the particle, this particle can arrive at r for the *first* time, thus yielding the D_{1r} domain as

$$\begin{aligned} D_{1r} &= \{(\epsilon_\alpha, l_\alpha) : \epsilon_\alpha \geq u^*_{\alpha r}(r),\, l_\alpha \leq l^*_{\alpha r}(r, \epsilon_\alpha)\} \,, \\ u^*_{\alpha r}(r) &\equiv \max_{r' \in S_r} u_\alpha(r') \,,\; l^*_{\alpha r}(r, \epsilon_\alpha) \equiv \min_{r' \in S_r} l_{\alpha r}(r', \epsilon_\alpha) \,, \\ S_r &= \{1 \leq r' \leq r \text{ for } \alpha = em \,,\; r \leq r' < \infty \text{ for } \alpha = i, e\} \,. \end{aligned} \tag{14}$$

Among the particles within this D_{1r} domain, some plasma particles (emitted electrons) can be *captured* by the probe (infinity). The ϵl-domain that contains all the *captured* particles is thus

$$\begin{aligned} D_0 &= \{(\epsilon_\alpha, l_\alpha) : \epsilon_\alpha \geq u^*_\alpha,\, l_\alpha \leq l^*_\alpha(\epsilon_\alpha)\} \,, \\ u^*_\alpha &\equiv \max_{r' \in S} u_\alpha(r') \,, \quad l^*_\alpha(\epsilon_\alpha) \equiv \min_{r' \in S} l_{\alpha r}(r', \epsilon_\alpha) \,, \quad S = \{1 \leq r' < \infty\} \,. \end{aligned} \tag{15}$$

Readily, the *reflected* sub-domain D_{2r} can be found as the relative complement of D_0 in D_{1r}, i.e., $D_{2r} = D_{1r} - D_0$.

After substituting $D_{2r} = D_{1r} - D_0$ into Eq. (12) and using Eqs. (14) and (15), the particle density becomes

$$n_\alpha(r) = \frac{2H_\alpha}{\pi} \int_{u^*_{\alpha r}(r)}^{\infty} \exp(-\epsilon_\alpha) \arcsin \frac{l^*_{\alpha r}(r, \epsilon_\alpha)}{l_{\alpha r}(r, \epsilon_\alpha)} \, \mathrm{d}\epsilon_\alpha - \frac{H_\alpha}{\pi} \int_{u^*_\alpha}^{\infty} \exp(-\epsilon_\alpha) \arcsin \frac{l^*_\alpha(\epsilon_\alpha)}{l_{\alpha r}(r, \epsilon_\alpha)} \, \mathrm{d}\epsilon_\alpha \,. \tag{16}$$

Such an integration does not only depend on the value of $\varphi(r)$ at the local r location, but also depends on the value at other r locations (as denoted by the star signs).

The integro-differential equation that describes the Vlasov-Poisson system is thus obtained by substituting Eq. (16) into Eq. (5). It is important to note that the solution of such a system depends only on five independent dimensionless parameters (δ_i, δ_p, φ_p, ρ_p, β), and yet it is independent of the mass ratio $\mu_i = m_i/m_e$.

To solve this integro-differential equation numerically, a non-uniform mesh for the spatial coordinate r is created and truncated at a maximum radius r_{max}. The values of the potential $\varphi(r)$ and density $n_\alpha(r)$ at all mesh points are contained inside two vectors $\boldsymbol{\varphi}$ and $\boldsymbol{n}_\alpha$. With an initial guess for $\boldsymbol{\varphi}$, the numerical scheme first computes the densities $\boldsymbol{n}_\alpha$ with Eq. (16) by carrying out the integrals numerically with Simpson's rule (as a Vlasov solver). Then, by using a finite-element formulation [12, 13], with the boundary conditions for a cylindrical probe as $\varphi|_{r=1} = \varphi_p$ and $\mathrm{d}\varphi/\,\mathrm{d}r|_{r_{max}} = -\varphi/r|_{r_{max}}$ [14], a Poisson solver for Eq. (5) calculates a new $\boldsymbol{\varphi}_{out}$ from the $\boldsymbol{n}_\alpha$ previously obtained. These two steps are the main building blocks of the Newton-Raphson iterative scheme that solves the non-linear algebraic equation $\boldsymbol{F}(\boldsymbol{\varphi}) = \boldsymbol{\varphi} - \boldsymbol{\varphi}_{out} = \boldsymbol{0}$. The most costly part of the algorithm is to numerically calculate the Jacobian for $\boldsymbol{F}$, which is carried out by several processors in parallel.

3. Current-voltage characteristics

Once $\varphi(r)$ is found, the current per unit probe length (normalized to the electron random thermal current I_{the}) can be found as

$$i_\alpha \equiv \frac{I_\alpha}{I_{the}} = \frac{2G_\alpha}{\sqrt{\pi}} \int_{u_\alpha^*}^{\infty} l_\alpha^*(\epsilon_\alpha) \exp(-\epsilon_\alpha)\, d\epsilon_\alpha\ , \quad I_{the} = 2\pi R_p e N_\infty \sqrt{\frac{k_B T_e}{2\pi m_e}} \tag{17}$$
$$G_i = -\sqrt{\delta_i/\mu_i}, \quad G_e = 1, \quad G_{em} = -2\beta\sqrt{\delta_p}\ ,$$

being invariant with r and positive for collected electron current. For a set of independent parameters (ρ_p, μ_i, δ_i, δ_p, β), the current-voltage characteristic ($i_t = i_e + i_i + i_{em}$ versus φ_p) can be obtained by varying the probe bias φ_p.

Based on the self-consistent full-kinetic model, the first results on the current-voltage characteristics were presented in [9] for an EP with radius $R_p \approx 3.6$ mm and immersed in an oxygen plasma with $T_i = T_e \approx 0.25$eV and $N_\infty \approx 1.07 \times 10^{12}\,\mathrm{m}^{-3}$. Together with the material thermionic properties $W \approx 2.5$eV and $\lambda_{RD} = 1$, the IV curves for different probe temperature T_p were calculated. Non-monotonic potential due to space-charge effects were successfully captured. This non-monotonic behaviour is not only present when the probe is negatively biased relative to the plasma ($\varphi_p < 0$), but also extends to positive probe bias ($\varphi_p > 0$). The higher the emission level is, this non-monotonic regime covers a larger range of φ_p. In this work, we consider the same ambient parameters as $\delta_i = 1$, $\mu_i = 29164.1$, $\delta_p = 0.32$, and $\beta = 1.716$, but varying the probe radius from $\rho_p = 0.1$ to $\rho_p = 2$ [see Fig. 1].

For a very negative probe bias ($\varphi_p \ll 0$), the probe emits electrons following the RD law, being invariant with φ_p. Because this emitted electron current is much larger than both the plasma ion current and the plasma electron current, the total current is also invariant with φ_p. The numerical results for such φ_p range can be found in [9]. Although not shown in Fig. 1, the IV curves for different ρ_p overlap with each other, as a result of normalizing the current with the thermal electron current. For a very large probe bias ($\varphi_p \gg 0$), the probe collects mainly electron current. This dimensionless current can depend on ρ_p if the probe collects electrons beyond the Orbital-Motion-Limited (OML) regime [11]. Nevertheless, numerical results are not shown for this φ_p range since it is not relevant for the purpose of this work.

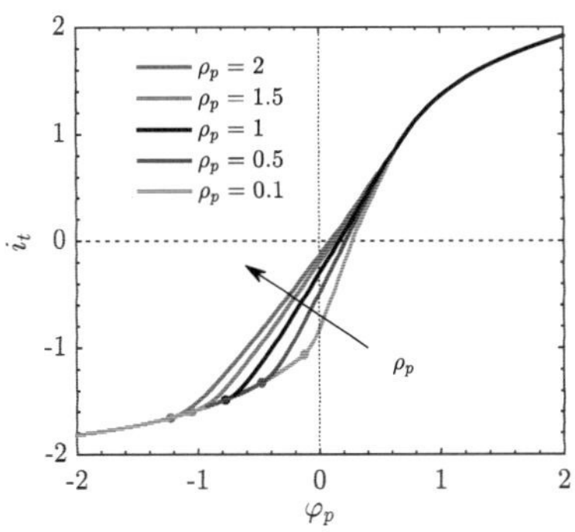

Figure 1. Current-voltage characteristics for different probe radius ρ_p, with $\delta_i = 1$, $\mu_i = 29164.1$, $\delta_p = 0.32$, and $\beta = 1.716$.

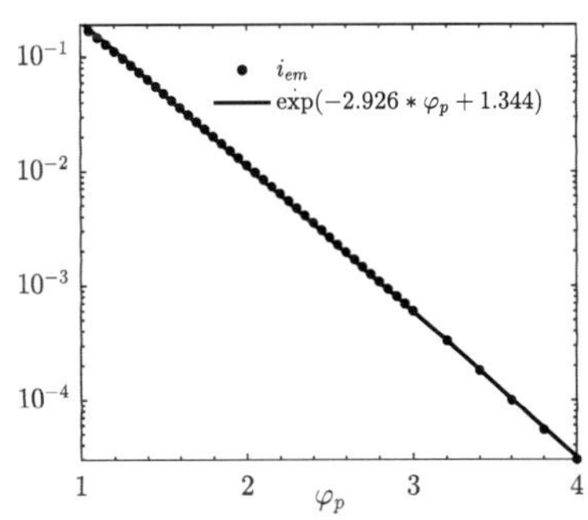

Figure 2. The calculated and fitted emitted electron current i_{em}, in the case of $\varphi_p > 0$ and monotonic potential profile ($\rho_p = 0.1$).

In Fig. 1, numerical results are presented for φ_p values that are close to zero. For negative probe bias ($\varphi_p < 0$), a dotted marker is placed on each curve to indicate the transition between the monotonic and non-monotonic regime. To the left of the transition point, the potential profile is monotonic. It can be observed that, as long as within the monotonic regime, all curves overlap with each other. Nevertheless, the net current no longer follows the RD law as being a constant value. It increases because the current from the plasma species becomes comparable with the emitted electron current. To the right of the transition point, the potential profile becomes no-monotonic and the emitted electron current is substantially reduced by the virtual cathode in front of the probe. The net current (positive for electron collection) thus rises rapidly and deviates from the overlapped route. For a larger probe radius, a larger net current implies more reduction of the emitted electron current, thus a stronger space-charge effect.

In the case of positive probe biases ($\varphi_p > 0$), for all the curves in Fig. 1, the non-monotonic potential profile persists for some positive probe biases ($\varphi_p > 0$). Because the RD current is larger than the random thermal current ($I_{RD}/I_{th} \approx 1.94$), the floating (zero-net-current) condition occurs at a positive probe bias. For $\varphi_p > 0$, the interest here is the emitted electron current [15]. This problem was first tackled by Langmuir. The electron current in *vacuum* was calculated considering electrons emitted from an internal cylindrical anode to an external coaxial cathodic cylinder, see Fig. 40 in [16]. According to the results from Langmuir, if i_{em} is plotted against φ_p with natural logarithm scale and linear scale respectively, the slope is estimated to be $1/\delta_p$. Our numerical calculations show that, as long as the potential profile is monotonic, i_{em} is identical for different ρ_p. In Fig. 2, the emitted electron current i_{em} (the dotted curve) for $\rho_p = 0.1$ is plotted (in decimal logarithm scale) versus φ_p (in linear scale). The value of i_{em} decreases exponentially, as indicated by the linear slope in this plot. After applying an exponential fitting (the solid line), the slope of the decreasing i_{em} in a natural logarithm scale is found to be 2.93, close to $1/\delta_p \approx 3.13$. The slight difference could result from the presence of plasma particles, which modifies the potential profile and thus the current.

3.1. *Inflection-point method*

The inflection point method makes use of the derivatives of the IV characteristics ($\mathrm{d}i_t/\mathrm{d}\varphi_p$) to measure the plasma potential. In [17], it is stated that, if an emissive probe is operated at sufficiently low emission level to minimize space-charge effects, the probe bias - at which the maximum of the derivatives occurs - is equal to the plasma potential. The inflection point refers to this maximum of the derivatives. Based on this theoretical prediction, in experiments, the inflection points are first measured for a number of low emission levels and then linearly extrapolated to obtain the inflection point for zero emission. Therefore, the plasma potential is given by the probe potential that coincides with this extrapolated inflection point in the limit of zero emission. The justification of this method has been provided qualitatively in [17] and quantitatively based on a model assuming planer emitter, cylindrical collector, and cold emitted electrons [15][18]. Before the attempts to seek for the justification by taking derivatives of the IV curves from numerical calculations, analytical results for zero emission can already provide significant insights.

If there is no emission ($\beta = 0$) and the probe is biased at the plasma potential ($\varphi_p = 0$), we first assume that there is no electric field anywhere and the plasma is directly connected to the probe without a sheath region. It can be proved later that such a flat potential is indeed a self-consistent solution of the Vlasov-Poisson system. In this case, there is no electric potential barrier that repels the plasma particles: $u^*_\alpha = u^*_{\alpha r} = u_{\alpha r} = 0$. Electrons and ions move towards the probe due to orbital effects alone. According to Eq. (13), the angular momentum limitation for the *non-forbidden* condition is given by ${l_{\alpha r}}^2 = r^2\epsilon_\alpha$. Because it decreases all the way towards the probe, we have the *sufficient* condition as $l^*_{\alpha r} = l_{\alpha r} = r\sqrt{\epsilon_\alpha}$ at every r. Consequently, the *captured* condition for the particles to be collected by the probe is determined by the *non-forbidden* condition at the probe only, i.e., $u^*_\alpha = u_{\alpha r}(r = 1)$ and $l^*_\alpha = l_{\alpha r}(r = 1)$. The current is then said to be orbital-motion-limited (OML) [19] [11]. Substituting $u^*_\alpha = 0$ and $l^*_\alpha = \sqrt{\epsilon_\alpha}$ into

Eq. (17), the currents are given by

$$i_i = -\sqrt{\delta_i/\mu_i}\ , \quad i_e = 1\ , \quad i_t = 1 - \sqrt{\delta_i/\mu_i}\ , \tag{18}$$

which coincide with the random thermal currents.

To prove that the flat potential profile is a self-consistent solution for a non-emitting probe biased at the plasma potential, the particle densities need to be found. By substituting $l^*_{\alpha r} = r\sqrt{\epsilon_\alpha}$ and $u^*_{\alpha r} = 0$ into Eq. (12), the electron and ion densities are given by [19]

$$n_i = n_e = 1 - \frac{\arcsin(1/r)}{\pi}\ , \tag{19}$$

which recovers the classical OML result with the particle density at the probe being 1/2. Equation (19) has been used as an initial condition in the analytical studies and the numerical simulations for transient effects [20] [21]. The reason for this normalized density to be less than one everywhere (i.e., particle density smaller than the plasma density) can be explained by Fig. 7 in [22]. Due to the particles that are captured by the probe, they do not turn back to the plasma, which results in a non-populated region in the distribution function for out-going particle populations. This non-populated region grows towards the probe. Because the particles arriving at the probe surface are collected by the probe (with no out-going particle populations), the distribution function becomes half-Maxwellian at the probe (thus $n_\alpha = 1/2$ under OML condition). According to Eq. (19), the net charge density is zero everywhere. Therefore, given by Poisson's equation, the flat potential profile is a self-consistent solution for $\varphi_p = 0$.

To determine the IV derivative near the plasma potential, it requires the expression for the net current i_t at the vicinity of $\varphi_p = 0$. It has been shown in the previous paragraphs that, if the probe is biased at the plasma potential ($\varphi_p = 0$), both ions and electrons currents do not only coincide with the thermal currents but also with the OML currents. Therefore, at the vicinity of $\varphi_p = 0$, the current for the attracted species can be expected to follow the OML law, which is confirmed by the numerical results [9][22]. For repelled species, the collection is also OML as the *captured* condition is determined by the *non-forbidden* condition at the probe only. Nevertheless, the electric potential barrier at the probe ($u^*_\alpha > 0$) leads to the Boltzman current law. By applying $u^*_\alpha = 0$ for attracted species, $u^*_\alpha = u_{\alpha r}(r=1)$ for repelled species, and $l^*_\alpha = \sqrt{\epsilon_\alpha + u_{\alpha r}(r=1)}$ for both, the currents at the vicinity of $\varphi_p = 0$ are exactly given by

$$\varphi_p < 0: \quad i_{e-} = \exp(\varphi_p)\ , \quad i_{i-} = -\sqrt{\delta_i/\mu_i}\left[\mathrm{erfcx}\left(\sqrt{-\varphi_p/\delta_i}\right) + 2\sqrt{-\varphi_p/(\pi\delta_i)}\right]\ , \tag{20}$$

$$\varphi_p > 0: \quad i_{e+} = \mathrm{erfcx}\left(\sqrt{\varphi_p}\right) + 2\sqrt{\varphi_p/\pi}\ , \qquad i_{i+} = -\sqrt{\delta_i/\mu_i}\exp(-\varphi_p/\delta_i)\ . \tag{21}$$

The derivatives of $i_{t-} = i_{e-} + i_{i-}$ and $i_{t+} = i_{e+} + i_{i+}$ are thus given by

$$\frac{\partial i_{t-}}{\partial \varphi_p} = \exp(\varphi_p) + \frac{\mathrm{erfcx}\left(\sqrt{-\varphi_p/\delta_i}\right)}{\sqrt{\delta_i\mu_i}}, \quad \frac{\partial i_{t+}}{\partial \varphi_p} = \mathrm{erfcx}\left(\sqrt{\varphi_p}\right) + \frac{\exp(-\varphi_p/\delta_i)}{\sqrt{\delta_i\mu_i}}\ . \tag{22}$$

For two different cases, the total current and the derivatives are plotted in Fig. 3. The inflection point of the IV curve (as the maximum of the derivative curve) at $\varphi_p = 0$ can be observed. Note that, due to $\mathrm{d}\ \mathrm{erfcx}(\sqrt{x})/\mathrm{d}\ x = \mathrm{erfcx}(\sqrt{x}) - 1/\sqrt{\pi x}$, the second derivative of the IV curve becomes to be positive infinity to the left of $\varphi_p = 0$ and negative infinity to the right.

As previously demonstrated, a flat-potential is a self-consistent solution for a non-emitting probe biased at the plasma potential. Since a flat potential requires an equal density of plasma ion and plasma electron everywhere [see Eqs. (5) and (19)], the presence of emitted electrons would thus violate this zero net charge and, according to Poisson's equation, a sheath region with non-zero electric field would build up. It can then be concluded that a flat potential can not be possible for an emissive probe biased at the plasma potential.

IOP Conf. Series: Journal of Physics: Conf. Series **958** (2018) 012001 doi:10.1088/1742-6596/958/1/012001

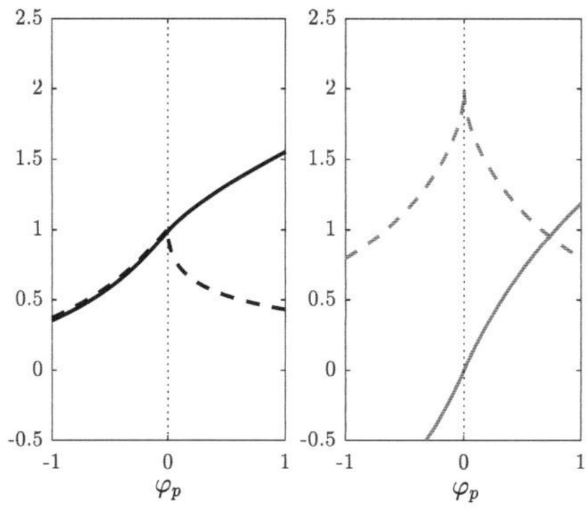

Figure 3. The i_t-versus-φ_p curve (solid curve) and the $\mathrm{d}i_t/\mathrm{d}\varphi_p$ derivative (dashed curve) near $\varphi_p = 0$ for two cases: $\delta_i = 1$ and $\mu_i = 1$ (left); $\delta_i = 1$ and $\mu_i = 10000$ (right).

4. Floating-potential characteristics

The probe is said to be floating if the net current is zero ($i_t = 0$). In laboratories, an emissive probe is heated to different temperatures (δ_p) to achieve different emission levels (β). The probe bias at the floating condition - the floating potential ($\varphi_f = e\phi_f/kT_e$) - is measured. The FP characteristics can thus be obtained by plotting φ_f versus δ_p (or versus other quantities that are equivalent to emission level, such as RD current and heating current). Such a FP curve was first shown by Kemp and Sellen, which rises rapidly at low emission and plateaus at high emission [6]. They claim that the knee (or break point) of this curve approximates the plasma potential. In 1967, by considering planar geometry and fluid approximation, Hobbs and Wesson found the monotonic to non-monotonic transition on a FP curve. The probe-to-plasma potential at that transition was found to be $e\phi_p \approx -k_B T_e$ (i.e., $\varphi_p \approx -1$) [3]. Since then, the floating-potential method is said to underestimate the plasma potential with an error being close to the electron temperature T_e.

To calculate this FP characteristic (φ_p versus δ_p), our numerical scheme is modified. The vector function $\boldsymbol{F} = \boldsymbol{0}$ (solved by the Newton-Raphson iterative scheme) considers the floating probe bias φ_f and the floating condition $i_t = 0$ as an additional unknown and an additional equation, respectively. The parameters that determine the FP characteristics are (ρ_p, μ_i, δ_i, δ_s, δ_w). The emission level β (a parameter in the Vlasov-Poisson system) is determined by δ_w and δ_s as given in Eq. (7).

In this work, the parameter set is first chosen as $\rho_p = 1$, $\mu_i = 29164.1$, $\delta_i = 1$, $\delta_w = 10$, and $\delta_s = 2 \times 10^{-10}$. The FP characteristics are then compared by varying one of the parameters as shown in Figs. 4-6. The probe is found to be able to float at positive probe bias ($\varphi_f > 0$), in agreement with the PIC simulation for planar geometry [5]. The saturated behaviour of the floating potential at high emission level is not observed for the parameters presented in this work. Such a non-saturating behaviour has also been found by PIC simulations for spheres [23] and in experiments performed for the plasma jet system [24]. For this reason, the knee of the FP curve is adopted in this work to evaluate the accuracy of floating-potential method. The knee is shown in Figs. 4-6 by a dot marker on each FP curve. To define this knee without ambiguity, it has been taken as the transition between non-monotonic and monotonic potential profile. That is to say, on each FP curve, to the left of the knee, the potential profile is monotonic (due to less emission), otherwise non-monotonic. If the knee occurs at $\varphi_f = 0$, the floating-potential method estimates the plasma potential precisely. If the knee occurs at $\varphi_f < 0$, the floating-potential method underestimates the plasma potential with the error being $|\varphi_f|$. What we are interested here is to see how this error varies with the ambient condition.

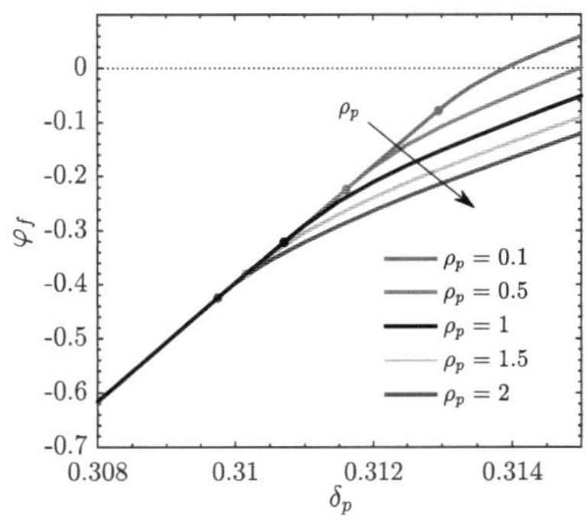

Figure 4. The floating-potential characteristics for different ρ_p.

The probe radius ρ_p is varied between 0.1 and 2, as shown in Fig. 4. The values of φ_f at the knee lie between 0 and -1, thus underestimating the plasma potential. They are also found to decrease with the probe radius. Nevertheless, this decreasing behaviour seems to weaken at larger radius as the curves become closer to each other. Although further calculations still need to be carried out to fully recover the planar solution $\varphi_f \approx -1$ for $\rho_p \gg 1$, this asymptotic behaviour suggests that this is viable. We can also say that, the smaller the probe radius is, the knee is closer to the plasma bias. Therefore, for a better approximation of the plasma potential using floating-potential method, a smaller probe radius is preferred.

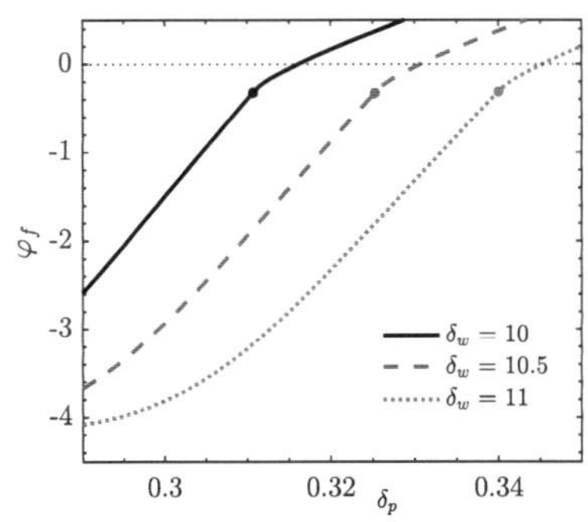

Figure 5. The floating-potential characteristics for different δ_w.

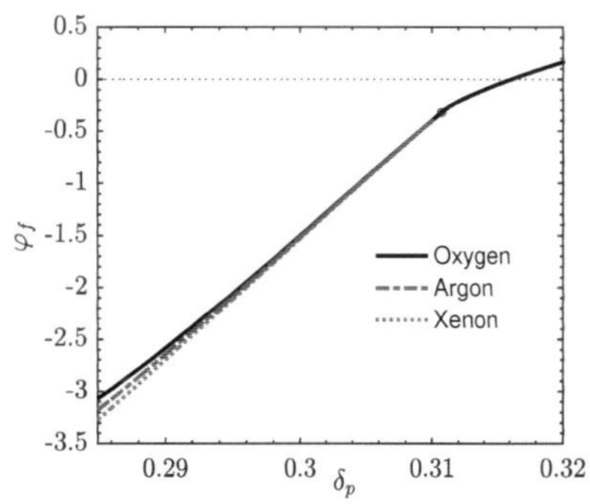

Figure 6. The floating-potential characteristics for different ion mass.

If the work function of the emitting material (δ_w) is reduced, the emission level (β) would increase according to Eq. (7). This noticeably modifies the floating-potential characteristics as shown in Fig. 5. If we look at the curves at a same probe temperature (δ_p), for a lower work function (δ_w), the floating potential (φ_f) increases due to higher emission level (β). Such a rise in φ_f serves to repel the excess of emitted electrons back to the probe and thus to attain the zero-net-current condition. Nevertheless, the influence of δ_w on φ_f weakens as δ_p decreases. This is because, for $\delta_p \ll \delta_w$, the change of the emission level β is dominated by δ_p other than δ_w [see Eq. (7)]. Although not shown in this figure, these three curves are superimposed at very low δ_p values, in that the floating potential is mainly determined by the plasma electron and ion current. At the end, since the knee is found to vary little with δ_w, the accuracy of the floating-potential method is the same for different materials.

Calculations were also carried out for Xenon and Argon [see Fig. 6]. Three curves are found to overlap with each other near the knee and for higher δ_p values, yet deviate from each other for low δ_p values. This results from the same reason mentioned in the paragraph above. Since the decrease in δ_p reduces the emission level β, the ion current becomes comparable to (or even can be larger than) the RD current. Under these circumstances, the floating potential would vary with the plasma-ion properties (such as ion mass) due to their influences on the ion current. If the ion mass is increased, the ion current tends to reduce. Therefore, to balance the plasma electron current and thus to retain the zero-net-current condition, the probe needs to float at a more negative potential to collect more ions, as shown by the dotted curve for Xenon. If the probe temperature δ_p increases, the floating potential also increases. When it becomes close to the plasma potential, as in the case of the knee in Fig. 6, the ion current becomes negligible compared to the emitted electron current. The influence of the ion properties on the knee thus becomes negligible and the knee is found to be independent on the ion mass. Due to this reason, calculations for different δ_i would not provide more information.

5. Conclusions

Based on the Orbital Motion Theory, a full-kinetic model for emissive probes with two-dimensional cylindrical geometry can solve the Vlasov-Poisson system self-consistently and determine the potential and density profiles for arbitrary plasma parameters (as long as the effects of collisions, plasma drift, particle trapping, transient effects, and magnetic fields are negligible) [9]. Such a model can capture the kinetic nature of the plasma sheath, reveal the non-negligible space-charge effects, solve for the non-monotonic potential profile without ambiguity, and unify both collecting (Langmuir) [19] and emissive probes in one compact framework. Some hypotheses for the numerical calculations presented this work can be easily relaxed within the framework of the model [22]: 1) the distribution function of the plasma and emitted species can be substituted by any energy distribution; 2) trapped particles can be included with a prescribed distribution function (under the condition of axial-symmetric electric field); 3) multiple plasma or emitted species, with arbitrary charge number (negative and positive), can be included.

Based on the numerical results, extensive parametric studies can be used to asses the accuracy of plasma-potential measurements using emissive-probe techniques. For the parameter range presented in this work, it is found that: (a) the space-charge effect increases with probe radius; (b) for a positive probe bias (relative to the plasma potential) and a monotonic radial potential profile, the emitted electron current decreases exponentially with the probe bias and the slope in the semi-natural-logarithmic plot is approximately $1/T_p$; (c) the probe can float at a positive bias relative to the plasma; (d) to determine the plasma potential using the knee of the floating-potential curve, a smaller probe radius is preferred; (e) the reliability of the floating-potential method is not influenced by the work function of the emitting material, neither the plasma-ion properties. An appropriate fitting of experimental EP measurements to this numerically calculated results can also be used to predict other plasma parameters, in addition to the plasma potential. Besides probe theory and plasma diagnostics, the numerical results can also benefit space applications such as spacecraft charging [25] and Low Work function Tethers (LWTs) for space debris removal [26].

To investigate the inflection-point method, analytical analysis for zero-emission was carried out, in the purpose to determine the current and its derivative for the probe bias at the vicinity of the plasma potential. Inflection point occurs exactly at the plasma potential, yet with the second derivative being infinite. In addition, for a non-emitting probe biased at the plasma potential, a flat potential is found to be a self-consistent solution of the Vlasov-Poisson system. In this case, the density of each species decreases towards the probe, yet with the net charge density being zero everywhere. If there is emission, emitted electrons result in negative space charge and a flat-potential solution is thus not self-consistent for an emissive probe biased at the plasma potential.

Acknowledgments
This work was supported by the Ministerio de Economía y Competitividad of Spain (Grant No ESP-2016-75887). Work by G. Snchez-Arriaga was supported by the Ministerio de Economía y Competitividad of Spain (Grant No RYC-2014-15357).

References

[1] Ionita C, Grnwald J, Maszl C, Starz R, Cercek M, Fonda B, Gyergyek T, Filipic G, Kovacic J, Silva C, Figueiredo H, Windisch T, Grulke O, Klinger T and Schrittwieser R 2011 *Contrib. Plasma Phys.* **51** 264–270
[2] Sheehan J P, Raitses Y, Hershkowitz N and McDonald M 2017 *J. Propul. Power* **33** 614–637
[3] Hobbs G D and Wesson J A 1967 *Plasma Physics* **9** 85
[4] Takamura S, Ohno N, Ye M and Kuwabara T 2004 *Contrib. Plasma Phys.* **44** 126–137
[5] Campanell M D and Umansky M V 2016 *Phys. Rev. Lett.* **116** 085003
[6] Kemp R F and Sellen J M 1966 *Rev. Sci. Instrum.* **37** 455–461
[7] Chang K W and Bienkowski G K 1970 *Phys. Fluids* **13** 902–920
[8] Schuss J J and Parker R R 1974 *J. Appl. Phys.* **45** 4778–4783
[9] Chen X and Sanchez-Arriaga G 2017 *Phys. Plasmas* **24** 023504
[10] Richardson O W 1916 *The emission of electricity from hot bodies* (Longmans, Green and co.)
[11] Sanmartin J R and Estes R D 1999 *Phys. Plasmas* **6** 395–405
[12] Choiniere E 2004 *Theory and experimental evaluation of a consistent steady-state kinetic model for 2-D conductive structures in ionospheric plasmas with application to bare electrodynamic tethers in space* Ph.D. thesis Univ. of Michigan
[13] Sanchez-Arriaga G and Sanmartin J R 2012 *Phys. Plasmas* **19** 063506
[14] Bernstein I B and Rabinowitz I N 1959 *Phys. Fluids* **2** 112–121
[15] Sheehan J P and Hershkowitz N 2011 *Plasma Sources Sci. Technol.* **20** 063001
[16] Langmuir I and Compton K T 1931 *Rev. Mod. Phys.* **3** 191–257
[17] Smith J R, Hershkowitz N and Coakley P 1979 *Rev. Sci. Instrum.* **50** 210–218
[18] Ye M Y and Takamura S 2000 *Phys. Plasmas* **7** 3457–3463
[19] Laframboise J G 1966 Theory of spherical and cylindrical langmuir probes in a collisionless, maxwellian plasma at rest Tech. rep. University of Toronto Institute for Aerospace Studies UTIAS Report, No. 100
[20] Sanmartin J R 1972 *Phys. Fluids* **15** 391–401
[21] Sanchez-Arriaga G 2013 *Phys. Plasmas* **20** 013504
[22] Chen X and Sanchez-Arriaga G 2017 *Phys. Plasmas* **24** 103515
[23] Delzanno G L and Tang X Z 2014 *Phys. Rev. Lett.* **113** 035002
[24] Marek A, Jlek M, Pickov I, Kudrna P, Tich M, Schrittwieser R and Ionita C 2008 *Contrib. Plasma Phys.* **48** 491–496
[25] Thiebault B, Hilgers A, Sasot E, Laakso H, Escoubet P, Genot V and Forest J 2004 *J. Geophys. Res. A* **109** A12207
[26] Sanchez-Arriaga G and Chen X *J. Propul. Power* Advance online publication. doi:10.2514/1.B36561

Multiprobe characterization of plasma flows for space propulsion

Julius Damba[1], P. Argente[1], P. E. Maldonado[2], A. Cervone[2], J. L. Domenech-Garret[1] and L. Conde[1]

[1] Departamento de Física Aplicada, ETSIAE, Univ. Politécnica de Madrid, 28040, Madrid, Spain

[2] Department of Space Engineering. Delft University of Technology. Kluyverweg 1, 2629HS Delft. The Netherlands

Abstract. Plasma engines for space propulsion generate plasma jets (also denominated plasma plumes) having supersonic ion groups with typical speeds in the order of tens of kilometers per second, which lies between electron and ion thermal speeds. Studies of the stationary plasma expansion process using a four-grid retarding field energy analyzer (RFEA), an emissive probe (EP) and a Langmuir probe (LP), all mounted on a three dimensionally (3D) displaced multiprobe structure are discussed. Specifically, the determination of plasma beam properties from the RFEA current–voltage (IV) characteristic curves is presented. The experimental results show the ion energy spectra to be essentially unchanged over 300 mm along the plasma-jet expansion axis of symmetry. The measured ion velocity distribution function (IVDF) results from the superposition of different ion groups and has two dominant populations: A low-energy group constituted of ions from the background plasma is produced by the interaction of the plasma jet with the walls of the vacuum chamber. The fast-ion population is composed of ions from the plasma beam moving at supersonic speeds with respect to the low-energy ions. The decreasing spatial profiles of the plasma-jet current density are compared with those of the low-energy ion group, which are not uniform along the axis of symmetry because of the small contributions from other ion populations with intermediate speeds.

E-mail: `julius.damba@upm.es`

1. Introduction

Electric propulsion (EP) has nowadays proved to be a growing technology for spacecraft propulsion. Rocket engines operate according to the basic principle of action and reaction, where the gas or plasma jet exhaust speed v_{ex} is crucial for momentum transfer. In conventional chemical thrusters, the maximum achievable gas speed is limited to the energy stored per unit mass of the propellant. EP systems overcome this limitation by use of supersonic plasma flows (also denominated *plasma plumes*) where ions accelerated by electromagnetic fields reach one to two orders-of-magnitude higher exhaust speeds. A higher specific impulse $I_{sp} = v_{ex}/g_o$, where g_o is the standard Earth acceleration, allows for a longer mission time and a heavier payload because of the reduced amount of required propellant [1, 2, 3, 4].

EP systems are characterized by large specific impulse and low thrust ($T = \dot{m}_i\, v_{ex}$) levels that rely on the mass ion production rate $\dot{m}_i$ achieved with the reduced electric power available on board. These features are essential for orbital maneuvres, station keeping, flight formation and/or end-of-life disposal of Earth-orbiting satellites. Today, hundreds of different thrusters are being successfully used for orbit correction as well as in deep-space scientific missions. Large telecommunication satellites are nowadays the main target consumers of high-power ion engines with typical thrust levels in hundreds of milli newtons and electric power consumption within the kilowatt range [1, 2].

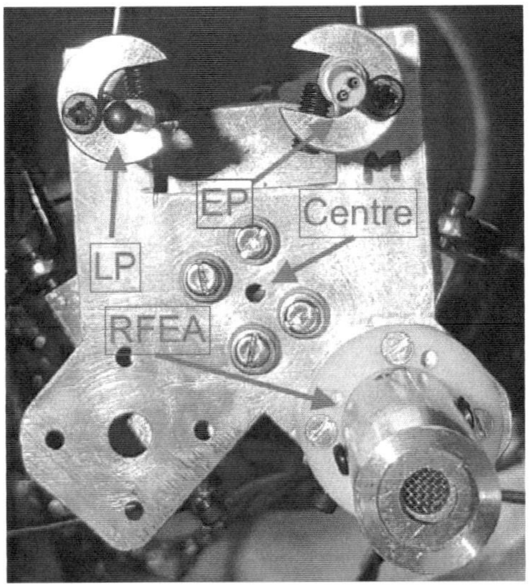

Figure 1. Front view of the multiprobe platform. The geometrical centre is indicated by the arrow.

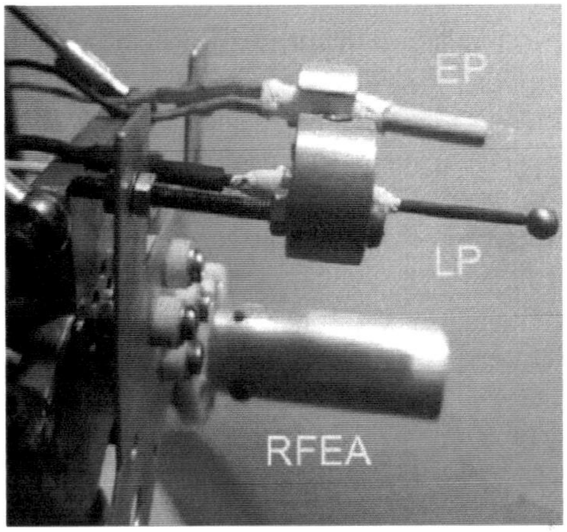

Figure 2. Side view of the LP, EP and RFEA probes mounted onto the platform, with their front ends aligned and facing the plasma thruster.

However, significant scientific and system engineering challenges are encountered during the design and integration of electric propulsion into small satellites (50–250 kg), such as the ones planned for constellations intended for global internet coverage and interactive television. These require thrust levels of roughly of 0.1–10 mN and electric power consumption below 500 W. Our alternative low power hybrid ion engine (ALPHIE) that has been the subject of a recent patent application falls within this category [5].

The outstanding characteristic of plasma thrusters for space propulsion is the acceleration of ions to supersonic speeds. This feature is investigated using the ion velocity distribution function (IVDF) of plasma streams in connection with the operational modes of the plasma thruster. The ALPHIE engine has been found to produce a mesothermal plasma flow [6] in which the exhaust plasma jet emerges with drift speeds that are intermediate between the ion and electron thermal speeds [7, 8, 9]. Such plasmas have been of great interest in recent studies of plasma flows, and are of crucial importance in electric propulsion and spacecraft–plasma interaction studies [10, 11, 12].

Also important but nevertheless not well understood characteristics are the spatial structure and physical mechanisms involved in the expansion of the plasma stream, as well as the interplays among different collisional processes (electron–neutral, ion–neutral, etc) and the ion energy relaxation mechanisms in the plasma stream. Understanding these would require pointwise measurements using a combination of different diagnostics, such as an emissive probe (EP), a Langmuir probe (LP) and a retarding field energy analyzer (RFEA) in order to obtain the spatial profiles of plasma properties like the electron density n_e, ion density n_i, plasma potential V_{sp} and the distribution of ion speeds.

In this paper we discuss the experimental results of the one-dimensional IVDF spatial profiles of mesothermal plasma flows from the ALPHIE thruster and the technical characteristics of the

IOP Conf. Series: Journal of Physics: Conf. Series **958** (2018) 012002 doi:10.1088/1742-6596/958/1/012002

computer-controlled multiprobe platform in Figs. (1) and (2). The structure of this paper is as follows: the features of the sliding positioning system are described in Sec. 2 and the electric probe diagnostics are discussed in Sec. 3. The experimental results are discussed in Sec. 4, specifically the two-peaked ion velocity spectra along the axis of symmetry of the plasma flow. Finally, we end in Sec. 5 with some concluding remarks.

2. The 3-D positioning/multiprobe system

The support structure for the positioning system fits exactly into our cylindrical stainless steel vacuum vessel of 80 cm length and 40 cm diameter, and it has the same cylindrical symmetry as the vacuum tank. The multiprobe platform that supports the measurement probes, shown in Figs. (1) and (2), is driven by three independent stepper motors.

The electric motors of the computer-controlled platform are shielded from any electromagnetic effects due to the plasma by carefully wrapping them in aluminium foils that are electrically connected to the grounded walls of the vacuum tank. Each stepper motor drives a rubber timing belt arranged inside an aluminium rail (also electrically grounded) that protects the belt from thermal loads and damage due to plasma exposure. This enables the setup to withstand long operation periods. The geometrical centre of the platform is indicated by an arrow in Fig. (1) and the combined effect of the three stepper motors produce 3-D rectangular displacements of this point with a spatial resolution better than ± 1 mm along the three spatial coordinates.

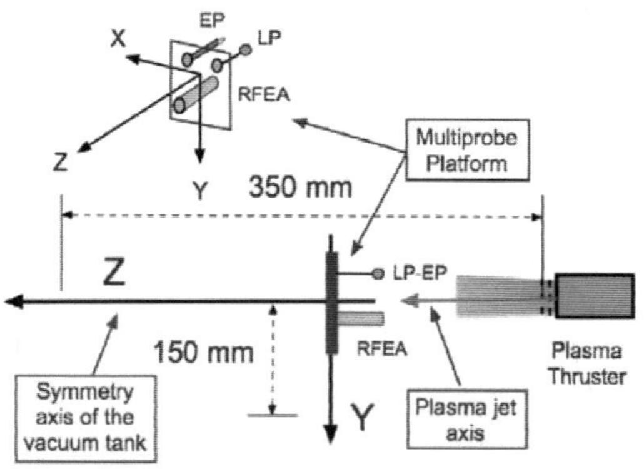

Figure 3. Schematic of the mechanical system for the multiprobe platform in Figs. (1) and (2). The probes face the plasma beam that is approximately centred at the Z-axis of symmetry.

The scheme in Fig. (3) shows the 3-D coordinate frame for the mechanical system, where the central point of Fig. (1) indicates the location of the Z axis, parallel to the symmetry axis of the vacuum tank. The mechanical system can displace the platform by 150 mm along both the X and Y directions and 300 mm along the Z coordinate axis.

As shown in Figs. (1) and (2), the probes were carefully and precisely mounted facing the plasma flow and aligned with a transversal plane to the platform. The RFEA is supported by screws whereas the LP and EP are mounted on D-shaped supports which can be released for maintenance and cleaning. However, a calibration is performed for the actual axial distances to account for the relative position of the exit section of the plasma thruster to the actual measuring range considered in Figs. (6), (7) and (8), i.e., 50– 350 mm. This ensures that the coordinates of each of the probes during measurements correspond to the real coordinates registered by the control software.

3. Plasma diagnostics

The plasma jet properties were characterized using a collecting Langmuir probe (LP) and a four-grid retarding-field energy analyzer (RFEA). The spherical stainless steel LP has a diameter of

4.44 mm and was biased with respect to the grounded walls of the vacuum tank using a time-sweep circuit.

The spherical probe was preferred to planar or cylindrical configurations to have minimum effect on the plasma jet expansion process. The probe was disposed as shown in Figs. (1–3), facing the plasma jet and its supporting shaft is oriented at the back side to minimize plasma wake effects on collected particles. Additionally, the deviation from spherical symmetry introduced by the electrically insulated support shaft of 1.33 mm diameter can be estimated using its cross sectional area. This correction is only 9% compared with the surface area of the Langmuir probe.

The LP probe was polarized by a 200 Hz frequency ramp signal added to a DC time-independent voltage [13]. The voltage–current (IV) characteristic curves were digitized by means of a Yokogawa DL9140 digital oscilloscope and stored for further analysis [13, 14, 15].

With the assumption of a Maxwellian energy distribution of electrons at 350 mm away from the plasma thruster, the typical electron densities in our experiments are in the range $n_e \simeq 0.1$–8.0×10^8 cm^{-3} with electron temperatures of $T_e = 1$–2 eV. These values give electron Debye lengths λ_{De} between 0.2 and 3.3 mm for the background plasma.

The RFEA [16, 17] operates under stationary conditions and has a 19.7 mm external diameter, with four grids that are equally spaced at $d = 5.0$ mm placed transversal to a cylindrical open channel. Ion motion in the analyzer is through this channel, which is 25.0 mm long with a cross section of 7.17 mm diameter. The ion current is collected by a copper plate located at the back end of the ion channel. The transparency of the system, T_r= 18.6%, was determined by optical methods.

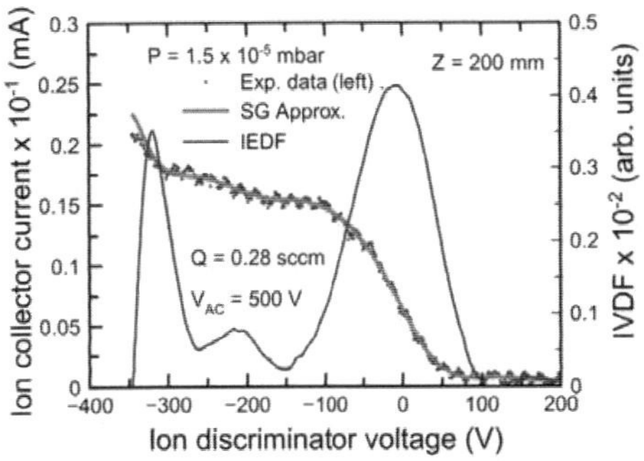

Figure 4. The original experimental data, a Savitzky-Golay numerical approximation of the IV curve (left axis) and the calculated IVDF (right axis) represented against V_{id} at $Z = 200$ mm.

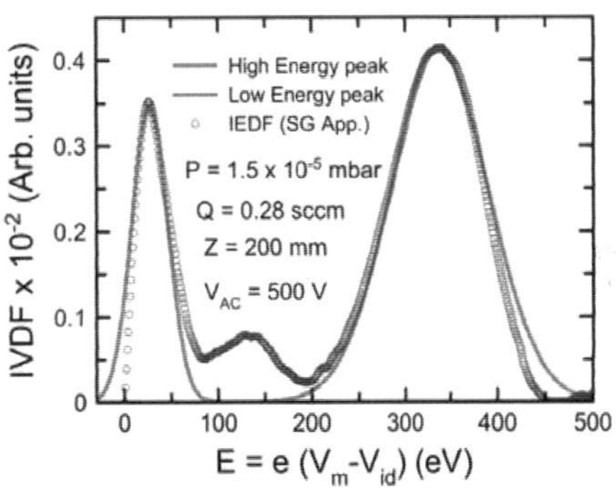

Figure 5. The IVDF (open symbols) of Fig. 4 and the two main peaks approximated by Gaussian functions (solid curves).

The RFEA voltage–current characteristic curves were obtained by measuring the ion-collector current as a function of the ion-discriminator grid voltage V_{id} relative to the (grounded) walls of the vacuum chamber. Such IV curves were digitized using a Keithley 2000 digital multimeter and a computer-controlled voltage source electrically connected to the ion discriminator grid. The typical electric potentials V_{es} and V_{er} of the secondary electron suppressor and the plasma electron repeller grids [16, 17], respectively, were $V_{es} \simeq V_{er}$= 25–30 V. These potentials repel the plasma electrons passing through the (electrically floating) RFEA first/front grid and suppress the secondary electron emission current produced by high-energy ions impacting the copper collector plate.

The ions with speeds $u_z > 0$ parallel to the axis of symmetry of the vacuum tank are collected and measured by the RFEA back plate. The ion energy resolution was about $\pm$ 8 eV for singly charged Argon ions of 400 eV energy. This is equivalent to radial speeds of ± 6 km s^{-1} for ions with axial speeds of 44 km s^{-1}.

The current density at the ion collector plate is $J_c = e\, u_z\, dn_i$, where $e > 0$ is the electronic charge; $dn_i = f(u_z)\, du_z$ is the number of ions with axial speeds between u_z and $u_z + du_z$ and $f(\boldsymbol{r}, u_z)$ is the stationary one-dimensional ion velocity distribution function at the point $\boldsymbol{r}$ where the RFEA is located. The ions reaching the collector plate have kinetic energies $E = m_i u_z^2/2 \geq eV_{id}$ to overcome the repelling potential V_{id} of the ion-discriminator grid. Hence, the ion current density collected by the RFEA copper plate is,

$$J_c(\boldsymbol{r}, u_{mz}) = e \int_{u_{mz}}^{\infty} u_z\, f(\boldsymbol{r}, u_z)\, du_z$$

Setting $dE = m_i\, u_z\, du_z$ and $s = m_i\, u_z^2/2$ we have,

$$J_c(\boldsymbol{r}, eV_{id}) = \frac{e}{m_i} \int_{eV_{id}}^{\infty} f(\boldsymbol{r}, s)\, ds \tag{1}$$

For a fixed position $\boldsymbol{r}$ the current density $J_c(\boldsymbol{r}, eV_{id})$ is a monotonically decreasing function of the discriminator voltage V_{id}. It falls from the maximum value when $V_{id} = V_m$, in which case all the ions can pass through the RFEA channel to impact the collector plate. The maximum current density $J_c(\boldsymbol{r}, eV_m)$ contains the contributions from all groups of ions in $f(\boldsymbol{r}, s)$ to the RFEA current–voltage curve. Then, $J_c(\boldsymbol{r}, eV_{id})$ in Eq. (1) decreases for $V_{id} > V_m$ because ions with low energies are rejected by the ion-discriminator grid and do not contribute to the total (integrated) current density.

The discriminator electric potential V_m that result in the maximum $J_c(\boldsymbol{r}, eV_m)$ corresponds to the lowest possible ion speed u_{mx} detected by the RFEA. The voltage V_m can be different from zero because the discriminator potentials V_{id} are not measured with respect to the local plasma potential but to the electric ground of the system. Setting $V_{id} = V_m$ in the lower limit of the integral in Eq. (1),

$$J_c(\boldsymbol{r}, eV_m) = e \int_0^{\infty} u_z\, f(\boldsymbol{r}, u_z)\, du_z = \frac{e}{m_i} \int_{eV_m}^{\infty} f(\boldsymbol{r}, s)\, ds \tag{2}$$

The maximum current density $J_{im}(\boldsymbol{r}) = J_c(\boldsymbol{r}, eV_m)$ can thus be calculated for each point $\boldsymbol{r}$ as,

$$J_{im}(\boldsymbol{r}) = \frac{I_m(\boldsymbol{r})}{T \times S}$$

where T_r is the transparency of the four-grid analyzer, S is the effective area of the copper collector plate and $I_m(\boldsymbol{r})$ is the maximum current of an RFEA current–voltage curve.

Furthermore, differentiating the ion current with respect to the bias potential of the ion-discriminator grid in Eq. (1) we have,

$$\frac{dJ_c}{dV_{id}} = -\frac{e^2}{m_i} f(\boldsymbol{r}, eV_{id}) \tag{3}$$

We make use of Eq. (3) in Sec. 4 to obtain the one-dimensional distribution function of the ion speeds. For this purpose, the IV curves of the RFEA were numerically approximated using the Savitzky-Golay (SG) scheme [18] to evaluate their derivatives, which are proportional to the one-dimensional IVDF in Eqs. (1) and (2).

The results of this analysis are shown in Figs. (4) and (5) where the original RFEA currents (blue dots) and their SG numerical approximation (solid curve) are represented against the bias potential of the ion discriminator grid. The shape of the IV curves indicates the presence of different ion groups in the one-dimensional IVDF (right axis), with maximum velocities realized for bias voltages where Eq. (3) has a maximum.

As shown in Fig. (5), the ion velocity distribution function can be represented against the scaled electric potential $e\,(V_m - V_{id}) = m_i u_z^2/2$ for $V_{id} > V_m$, so that $u_z \simeq 0$ for the maximum measured current $I_m(\boldsymbol{r})$ of the RFEA current–voltage curve. The IVDF in Fig. (5) has thus been plotted against the kinetic energy of ions, as well as in Fig. (6) where the energy axis is inverted for clarity.

The IVDF as in Fig. (5) has three peaks that can be considered to be the superposition of two or more ion populations, but with only two ion groups playing a dominant role. The SG approximations of the IV curves, as in Fig. (3), along the Z axis are represented in Fig. (6). Finally, the ion temperatures can be evaluated using the full width at half maximum (FWHM) values of the Gaussian fits to the two main peaks of the IVDF as in Fig. (5).

4. Experimental results

The limited pumping speed of our vacuum equipment determines the maximum possible Argon mass flow rate for the operation of the ALPHIE plasma thruster, and consequently its ion production rate $\dot{m}_i$. However, sufficient plasma production in Argon is achieved with neutral gas flow rates as low as $Q = 0.28$ sccm (standard cubic centimeter by minute), allowing for a low gas pressure of $p_a = 1.5 \times 10^{-5}$ mbar to be achieved inside the plasma chamber as shown in Figs. (6), (7) and (8).

The multiprobe stand was employed for the precise positioning of the probes at different points $\boldsymbol{r}$ within the mesothermal plasma flow generated by our ALPHIE plasma thruster that operates under stationary conditions. Supersonic ion streams with typical speeds in the range 37–44 km s^{-1} which are higher than the characteristic ion sound velocities c_{is} =10–11 km s^{-1} at a fixed point 20 cm from the thruster exit section have been observed in previous studies [6].

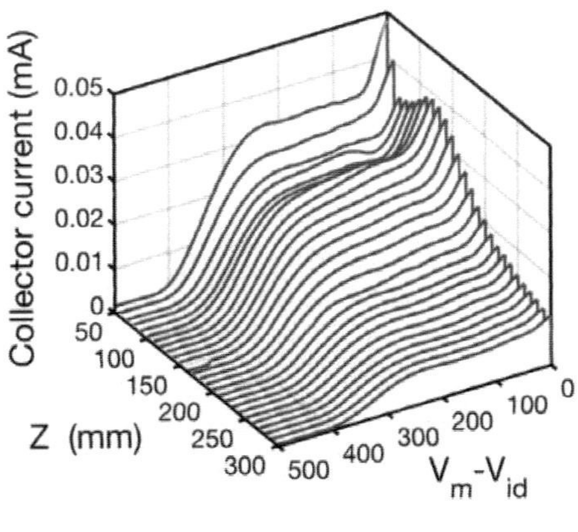

Figure 6. A waterfall representation of SG approximations of the RFEA current–voltage characteristic curves along the Z axis of the scheme in Fig.(3).

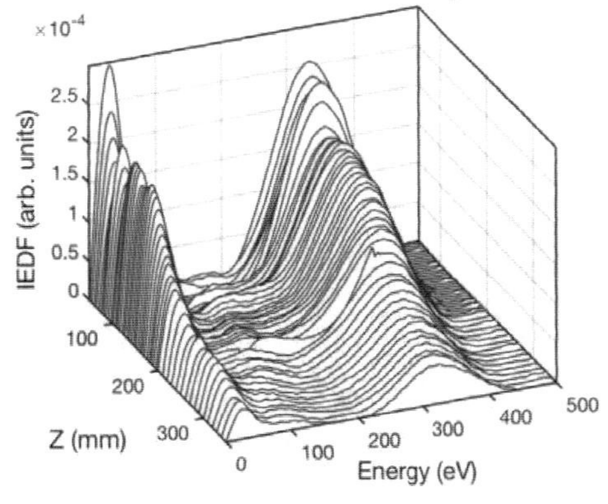

Figure 7. Waterfall representation of the two-peaked IVDFs, calculated from the IV curves of Fig. (6), along the Z axis against the kinetic energy.

These general characteristics are confirmed by the waterfall representation (see Fig. 7) of the two-peaked IVDFs of the scheme of Fig. (3) along Z which is approximately aligned with the centreline of the plasma jet. The same distribution functions are plotted against the kinetic energy $E = e\,(V_m - V_{id})$, as shown in Fig. (5).

The maxima of the fast ion group are around 330 eV and at about 30 eV for the low-energy populations. The speeds corresponding to the low-energy ion group peaks remain essentially independent of the operational parameters of the ALPHIE thruster, whereas the velocity of the fast ions depends on the acceleration voltage [6] (which was fixed at $V_{ac} = 500$ V in this case). The velocities in Fig. (7) remain essentially constant over 50–350 mm along the axial direction. Additionally, the height of the ion velocity distribution function decreases with the radial coordinate Z as the plasma jet expands.

The characteristic cross sections for collision (elastic and resonant charge exchange) of neutral atoms with Argon ions are in the order of $\sigma_{ia} \sim 3 \times 10^{-15}$ cm^2 [19, 20], which correspond to mean free path of $\lambda_{ai} \simeq 500$ cm for our working pressures. We also obtain $\lambda_{aa} \simeq 430$ cm for elastic collisions between two neutral Argon atoms [19]. The cross sections for electron and ion impact ionization of Argon also lead to mean free paths exceeding the size of our vacuum tank [20]. Furthermore, the plasma potential variations measured with the emissive probe along the axial direction are in the order of a few volts, well below the characteristic kinetic energy of the ions in Fig. (7). Thus, there is no evidence for axial electric fields that could appreciably change the velocity of heavy particles in the plasma.

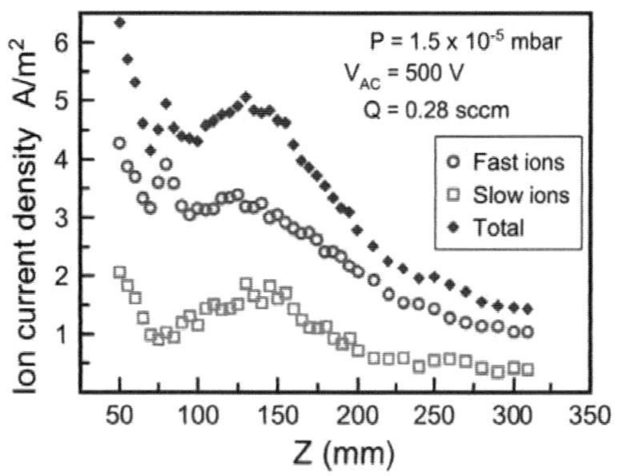

Figure 8. Current densities along the Z axis of Fig. (3), calculated from the RFEA current–voltage curves in Fig. (6). The open circles and squares show, respectively, the current densities $J_{if}(z)$ and $J_{il}(z)$ of the ions with kinetic energies above and below 200 eV. The solid diamond shapes represent the sum of the two ion group contributions.

The plasma thruster is the only source of ions in our experiment, and therefore the low-energy ion groups in Figs. (4) and (7) are due to the interaction of fast positively charged particles with the vacuum chamber walls. A number of ions from the plasma beam recombine after colliding with the walls, but a significant proportion is reflected back towards the plasma chamber after losing a large fraction of their initial energy. The resulting heavy particles, with reduced speeds, must be responsible for the formation of the low-energy peak in the IVDF. Hence, the speeds of ions in this low-energy group must be in random directions yet the ion velocity distributions in Fig. (7) only account for those with $v_z > 0$ present at the point where the RFEA is located. On the contrary, fast ions have a very important proportion of velocity components along the direction of the plasma jet expansion, parallel to the Z axis in Fig. (3).

The FWHM estimates of ion temperatures for the two main peaks in Figs. (5) and (7) give similar values along the Z axis. For the low-energy ion group we obtain $k_B T_{il} = 30$ eV, whereas $k_B T_{if} = 120$ eV for the fast-ion population. Both ion temperatures are higher than the typical electron temperatures of the distant background plasma. The kinetic energy range 200–400 eV of the fast-ion group peaks in Fig. (7) leads to ion velocities of 31–44 km s^{-1}, which are much higher than the ion sound speed $c_{is} = \sqrt{k_B(T_i + T_e)/m_i} \sim 8.5$ km s^{-1}. The average supersonic drift speed $u_D = u_{df} - u_{dl} \simeq 38$ km s^{-1} of the fast ion group can be estimated by the difference between the average peak velocities $u_{dl} \simeq 15$ km s^{-1} and $u_{df} \simeq 53$ km s^{-1} of the slow and fast ion groups, respectively.

The axial profile of the stationary plasma jet expansion along the axis of symmetry (Z) is shown in Fig. (8), where the maximum RFEA ion current densities $J_{im}(z)$ are referred to. These values are calculated from the current maxima $I_m(z)$ in Fig. (6) and they account for the contribution to the RFEA ion current of all ion populations in the energy spectra of Fig. (7). Unfortunately, the charged-particle densities in the plasma beam cannot be estimated from the current densities in Fig. (8) because of the broad energy range in Fig. (7).

However, the current density $J_c(z, eV)$ in Eq. (2) is only due to the contribution of ions with energies $E \geq eV$ to the RFEA ion current and we make use of this fact to estimate the axial current density of ions in the plasma jet. Fig. (7) shows the two main IVDF peaks to be separated at energy $E_l \simeq 200$ eV. Therefore, $J_c(z, E_l) = j_{if}(z)$ is approximately the current density $j_{if}(z)$ of the plasma-beam ions whereas $J_{il}(z) = J_{im}(z) - J_{if}(z)$ is the contribution of low-energy ions in the IVDF to the RFEA maximum current. The values of $j_{if}(z)$ (open circles) and $J_{il}(z)$ (open squares) are also represented together with $J_{im}(z)$ (solid diamonds) in Fig. (8).

The plasma beam ion current density $J_{if}(z)$ decreases by a factor 2.7 over 300 mm along the axial direction Z that is collinear with the central direction of propagation of the plasma jet. The flow of supersonic ions is maximum close to the exit section of the plasma thruster and decreases along the axial direction. The flow due to low-energy ions $J_{il}(z)$ is spatially irregular, always less than $J_{if}(z)$ and is maximum around 150 mm. The increase in both $J_{if}(z)$ and $J_{il}(z)$ around $Z = 150$ mm results from the contribution to the measured RFEA ion current by the small peak between 100–200 eV, which can be observed in Figs. (5) and (7). The estimation of the $J_{if}(z)$ and $J_{ij}(z)$ was made by simply setting $E_l = 200$ eV which divides the contribution of this small third ion group between both current densities.

5. Conclusions

We have been able to show how a combination of the three diagnostics on the movable multiprobe platform constitutes a valuable tool to explore the plasma expansion process for supersonic plasma jets. Plasma beam diagnostics requires a combination of different probes since the RFEA which is the conventional tool for plasma energy analysis has a directional character. Only heavy charged particles with speeds parallel to RFEA axis are detected at its end collector plate. Therefore, the displacement of the RFEA axis should reflect the symmetry of the plasma expansion process. In view of this, the RFEA diagnostic needs to be complemented with emissive and Langmuir probes to obtain a complete physical picture of the plasma expansion process.

Here we focused our attention on the determination of plasma properties along the Z axis shown in the schematic of Fig. (3). This axial direction represents the symmetry of collimation of the hemispherical expansion of the plasma jet, allowing reliable measurements of the IVDF with the RFEA. The velocity distribution functions of ions in Fig. (7) confirm the general characteristics of the mesothermal plasma jets produced by the ALPHIE plasma thruster previously observed [6].

As we have shown, two ion groups dominate the expansion of ALPHIE plasma jets along the axial direction. The preserved ion velocity distribution functions in Fig. (7) along the Z axis indicate how the supersonic ion group remains essentially unchanged and is only modified by the interaction with the vacuum chamber walls. The low-energy peak in the IVDF corresponds to ions reflected back from the walls with lower energies. Unfortunately, we are not in position to explain the physical origin of the additional low peaks in Fig. (7) at intermediate energies of 100–200 eV that appear at some points. The ion energy losses by collisional energy exchange are excluded because of the long mean free paths required.

Future work will address the determination of plasma jet radial dependence of the IVDF and their influence on beam collimation, as well as the study of the electron energy distribution functions.

Acknowledgments
The authors are indebted to Dr. J. González for his useful suggestions. The technical assistance from Mr. F. Sánchez is also appreciated. This work was funded by the Ministerio de Economía, Ciencia e Innnovación of Spain under Grant ESP2013–41078-R. The authors also acknowledge the financial support from Aernnova Aerospace SAU.

References
[1] Goebel D M and Katz I 2008 *Fundamentals of electric propulsion. Ion and Hall thrusters* (John Wiley & Sons, Hoboken, New Jersey, USA) Chaps. 1, 6 and 7
[2] Mazouffre S 2016 *Plasma Sources Sci. Technol.* **25** 033002–1,27
[3] Charles C 2009 *J. Phys. D: Applied Phys.* **42** 163001–1,18
[4] Ahedo E 2011 *Plasma Phys. Control. Fusion* **53** 124037–1,18
[5] Conde L, Domenech-Garret J L, Donoso J M, Del Río E and Castillo M A 2015 Plasma accelerator with modulated thrust. Patent application PCT/EP2015/074879. European Patent Office (2015)
[6] Conde L, Domenech-Garret J L, Donoso J, Damba J, Tierno S P, Alamillo-Gamboa E and Castillo M A 2017 Supersonic plasma beams with controlled speed generated by the alternative low power hybrid ion engine (ALPHIE) for space propulsion.
[7] Zhang Z, Tang H, Zhang Z, Wang J and Cao S 2016 *Rev. Sci. Instrum.* **87** 123510–1,8
[8] Hu Y and Wang J 2015 *IEEE Transactions on Plasma Science* **43** 2832–2838
[9] Merino M and Ahedo E 2015 *IEEE Trans. on Plasma Sci.* **43** 244–251
[10] Charles C and Boswell R W 2004 *Phys. Plasmas* **11** 1706–1714
[11] Charles C, Boswell B W and Lieberman M A 2006 *Appl. Phys. Lett.* **89** 261503–1,3
[12] Corr C S, Zanger J, Boswell B W and Charles C 2007 *Appl. Phys. Lett.* **91** 241501–1,3
[13] Troll O, Conde L, Criado E, Donoso J M and Herdrich G 2010 *Contrib. Plasma Phys.* **50** 819–823
[14] Tierno S P, Domenech-Garret J L, Donoso J M, Jennewein D, Herdrich G, Fasoulas S and Conde L 2013 *IEEE Trans. on Plasma Sci.* **41** 695–700
[15] Hershkowitz N 1989 *Plasma Diagnostics. Discharge Parameters and Chemistry.* vol 1 ed Auciello O and Flamm D L (Academic Press, San Diego, CA, USA) chap 3, pp 113–182
[16] Böhm C and Perrin J 1993 *Rev. of Sci. Instrum.* **64** 31–44
[17] Hutchinson I H 1990 *Principles of plasma diagnostics.* (Cambridge University Press, New York, USA)
[18] Magnus F and Gudmundsson J T 2008 *Rev. Sci. Instrum* **79** 073503–1,8
[19] Raizer Y P 1991 *Gas Discharge Physics* (Springer Verlag, Berlin, Germany)
[20] Lieberman M A and Lichtenberg A J 1994 *Principles of Plasma Discharges and Materials Processing.* (John Wiley & Sons,, New York, USA)

IOP Conf. Series: Journal of Physics: Conf. Series **958** (2018) 012003 doi:10.1088/1742-6596/958/1/012003

Feasibility, strategy, methodology, and analysis of probe measurements in plasma under high gas pressure

V I Demidov[1], M E Koepke[1], I P Kurlyandskaya[2] and M A Malkov[2]

[1] Department of Physics and Astronomy, West Virginia University, Morgantown, WV 26506, USA
[2] INTEPH Technologies LLC, OH 45066, USA

E-mail: vladimir.demidov@mail.wvu.edu

Abstract. This paper reviews existing theories for interpreting probe measurements of electron distribution functions (EDF) at high gas pressure when collisions of electrons with atoms and/or molecules near the probe are pervasive. An explanation of whether or not the measurements are realizable and reliable, an enumeration of the most common sources of measurement error, and an outline of proper probe-experiment design elements that inherently limit or avoid error is presented. Additionally, we describe recent expanded plasma-condition compatibility for EDF measurement, including in applications of large wall probe plasma diagnostics. This summary of the authors' experiences gained over decades of practicing and developing probe diagnostics is intended to inform, guide, suggest, and detail the advantages and disadvantages of probe application in plasma research.

1. Introduction

It is known [1] that, in a low-pressure plasma when there are no collisions between atoms and electrons flowing to the electric probe, the probe electron current I_e may be calculated using the Langmuir formula

$$I_e(V) = \frac{2\pi N_e eS}{m^2}\int_{eV}^{\infty} f(\varepsilon)(\varepsilon - eV)\mathrm{d}\varepsilon, \qquad (1)$$

where $f(\varepsilon)$ is the electron distribution function (EDF), N_e is the electron density, e is the magnitude of electron charge, S is the probe surface, m is the mass of electron, V is the (negative) probe potential and ε is the electron energy. In equation (1), it is assumed that

$$4\sqrt{2}\pi m^{-3/2}\int_{0}^{\infty}\sqrt{\varepsilon} f(\varepsilon)\mathrm{d}\varepsilon = 1.$$

At present, the most published probe measurements of the EDF [2-6] are conducted at low gas pressure and interpreted with the Druyvesteyn formula, which is simple to obtain from equation (1) by the double differentiation over the probe potential [7]

$$f(\varepsilon) = \frac{m^2}{2\pi N_e S e^3} \times \frac{d^2 I_e}{dV^2}. \qquad (2)$$

The method for finding the plasma parameters by applying equations (1) and (2) is limited by the important requirements that the mean-free-path of electrons λ is much greater than both the Debye shielding radius r_D and the probe radius a. It is still generally accepted that, due to violation of those conditions, the probe method is not applicable (in noble gases) at pressures of greater than, typically, 10 Torr. For advancing the probe method into the region of high pressures in order to find the EDF, a number of theories have been proposed. In this paper, we briefly review the available theories and analysis of a number of experimental studies on the EDF measurements and electron densities in high (medium) pressure plasma.

2. The effect of sink of electrons to the probe

The effect of the electron sink to the probe is connected to reducing electron density near the probe due to electron absorption (recombination) by the probe surface. The distortion of the EDF occurs due to the finiteness of the rate of diffusion replenishment of the region of electrons depleted due to their flow to the probe. This reduces the probe current at any energy. In the sink theory, it is assumed that at distances from the probe surface starting at the infinity to the mean free path of electrons, λ, electrons flow to the probe due to diffusion and from distance λ to the probe surface they go to the probe freely without collisions. It is also assumed that the mean-free-path of electrons and the probe radius are much greater than r_D.

The first was Swift's calculation [8], where for a spherical probe the variation of the EDF, $f(\varepsilon, \lambda)$, at the boundary of the collisionless region from the undisturbed EDF, $f(\varepsilon)$, was modeled. For the dependence of the electron current on the retarding potential, expression

$$I_e(V) = \frac{2\pi N_e e S}{m^2} \int_{eV}^{\infty} \frac{f(\varepsilon)(\varepsilon - eV)d\varepsilon}{1+\delta(\varepsilon)\left(1-\frac{eV}{\varepsilon}\right)} \tag{3}$$

was obtained. Here,

$$\delta(\varepsilon) = \frac{3a^2}{4\lambda(\varepsilon)(a+\lambda(\varepsilon))} \tag{4}$$

is called as the sink parameter (of electrons to the probe). The influence of this effect on the standard (that is, when equation (2) is valid for the EDF) probe measurements of space potential, plasma density, and the average energy (temperature) of the electrons has been investigated, for example, in Refs. [8-10].

At present, in the experimental practice of probe measurements, the cylindrical probes are used much more often than spherical ones, since they are the simplest in manufacturing and the influence of the probe holder on measurements is minimal. The calculations in Ref. [9], resembling the are calculations from Ref. [8] but appropriate for the current to a cylindrical probe having length l and radius a, also give equation (3) but with the sink parameter

$$\delta(\varepsilon) = [3a/4\lambda(\varepsilon)] \ln\left[1 + \frac{2l}{a+\lambda}\right].$$

Note that, in the last case, the electron diffusion flux was taken into account not only on the side surface of the probe, but also, on one probe end. If we take into account the diffusion of electrons to both ends of the probe in equation (3), we obtain

$$\delta(\varepsilon) = \frac{3a}{4\lambda(\epsilon)} \ln\left[1 + \frac{l}{a+\lambda}\right]. \tag{5}$$

A generalized theory for the sink effect of electrons on a probe of arbitrary shape is also possible. Such a generalization has been conducted, for example, in Ref. [11], where the electron current was determined for a probe made in the form of an ellipsoid of revolution with semi-axes of $l/2$ and a. Using the coordinates of an elongated or an oblate ellipsoid of revolution as a result of calculations similar to Ref. [8-10], the expression for the current to the probe also yields expression (3), where the sink parameters are:

a) for a prolate ellipsoid of revolution ($l/2 > a$)

$$\delta(\varepsilon) = \frac{3S}{32\pi b\lambda(\epsilon)}\ln\left(\frac{\sigma_0+1}{\sigma_0-1}\right);\ b = \sqrt{(\frac{l}{2}+\lambda)^2-(a+\lambda)^2};\ \sigma_0 = \frac{\frac{l}{2}+\lambda}{b}; \tag{6}$$

b) for an oblate ellipsoid of revolution ($l/2 < a$)

$$\delta(\varepsilon) = \frac{3S}{32\pi b\lambda(\epsilon)}[\pi - 2\tan^{-1}\sigma_0];\ b = \sqrt{(a+\lambda)^2-(l/2+\lambda)^2};\ \sigma_0 = \frac{\frac{l}{2}+\lambda}{b}. \tag{7}$$

Analysis of expressions (6) and (7) shows that for $l/2 = a$, that is for the case of a spherical probe, they give expression (4). If, however, an ellipsoid with semi-axes $l/2$ and a is assumed to be an equivalent of a cylindrical probe for which $l >> a$, then for the value of $\delta(\varepsilon)$ in equation (6) for the transition with limits $\sigma_0 \to 1$ and $a \geq \lambda$

$$\delta(\varepsilon) \approx \frac{3a}{4\lambda}\ln\left[\frac{l}{a+\lambda}\right], \tag{8}$$

which practically coincides with equation (5).

So, as shown by the available calculations, the dependence of the electron current to the probe of arbitrary shape with the retarding potential (in the framework of the theory of the sink effect) can be described by unique equation (3), where the sink parameter is determined by the probe geometry and calculated either by equation (6) or equation (7), respectively, for a prolate or oblate probe. It is easy to see from equation (3) that, in the Langmuir mode of operation ($a/\lambda << 1$), the current on the probe is determined by equation (1). For the opposite case ($a/\lambda >> 1$) the sink theory gives

$$I_e(V) = \frac{2\pi N_e e S}{m^2}\int_{eV}^{\infty}\frac{f(\varepsilon)\varepsilon d\varepsilon}{\delta(\varepsilon)}. \tag{9}$$

From equation (9) by the differentiation over probe potential [12] it is simple to obtain

$$f(\varepsilon) = \frac{\delta(eV)m^2}{2\pi N_e S V e^3} \times \frac{dI}{dV}, \tag{10}$$

which is an analogue of the Druyvesteyn formula for the collisional plasma.

3. Kinetic probe theory

Within the framework of the sink theory, the influence of collisions at the distances less than λ from the probe surface on the electron current is neglected. But, strictly speaking, some of the electrons streaming toward the probe from a distance λ can be prevented from reaching its surface by collisional encounters with gas atoms. To develop more precise predictions, it is necessary to take into account those collisions and to choose realistic boundary conditions on the surface of the probe. For this purpose, a kinetic theory can be used.

3.1. Diffusion mode ($\lambda \ll a \ll \lambda_\varepsilon$)

The kinetic theory for this case in the diffusion mode, that is $a/\lambda >> 1$, has been studied in Ref. [12]. Since electrons undergo many collisions before they reach the probe, the EDF near the probe is weakly anisotropic and, hence, the two-term expansion is valid [5,6]. In a stationary weakly ionized plasma, the isotropic part of the EDF, neglecting inelastic collisions, is described by the kinetic equation

$$\frac{v}{3}div\,\vec{f_1} - \frac{eE}{3m}\frac{1}{v^2}\frac{\partial}{\partial v}\left(v^2\vec{f_1}\right) = \frac{m}{M}\frac{1}{v^2}\frac{\partial}{\partial v}(v^3 f_0 \nu_{en}), \tag{11}$$

where ν_{en} is the frequency of transport-inducing electron collisions with gas atoms, and $\vec{f_1}$ is the directed part of the EDF, which may be described by equation [13]

$$v\nabla f_0 - \frac{e\vec{E}}{m}\frac{\partial f_0}{\partial v} = -\nu_{en}\vec{f_1}. \tag{12}$$

If inequality $\overline{a} = a + d \ll \lambda_\varepsilon$ holds, the EDF is nonlocal [12] and it is possible to neglect the right-hand side in equation (11), which is responsible for energy losses in elastic collisions and is of the order of $(m/M)f_0\nu_{en}$ in comparison with the left-hand side, which is of the order of $v^2 f_0/(\nu_{en}\overline{a})^2$. Here, $\overline{a}$ is the characteristic scale of the EDF variation near the probe and d is the probe sheath/pre-sheath size. In this case, it is convenient [12] to use as an independent variable w instead of ε, where $w = \varepsilon + e\varphi(r)$ is the total energy of an electron with the kinetic energy ε and $\varphi(r)$ is the plasma electric potential at distance r. After the calculation of the isotropic part of the EDF, we obtain

$$f_0(w,r) = C_0(w) + C_1(w)\int \frac{dr\, \nu_{en}(v)}{v^3 r^\alpha}, \tag{13}$$

where $C_0(w)$ and $C_1(w)$ are functions of the total electron energy, $\alpha = 1$ and $\alpha = 2$ are for cylindrical and spherical probes, respectively.

Function $C_0(w)$ may be found from the condition that at infinity the unknown distribution function is equal to the undistorted EDF. Assuming that potential φ at the infinity satisfies the requirement $\varphi(\infty) = 0$, we find that $C_0(w) = f(w)$. The second boundary condition is $f(\varepsilon \geq eV, a) = 0$. Therefore, it is assumed that the density of "energetic" electrons having kinetic energy greater than eV on the probe surface is zero. Then, it is found that the distribution function becomes

$$\begin{cases} f_0(w,r) = f_0(w) & \text{for } w \leq eV, \\ f_0(w,r) = f_0(w)\left(1 - \int_r^A \frac{dr\, \nu_{en}}{r^\alpha v^3} \Big/ \int_a^A \frac{dr\, \nu_{en}}{r^\alpha v^3}\right) & \text{for } w \geq eV. \end{cases} \tag{14}$$

For the case of the cylindrical probe in equations (14), the upper limit of integration A may be taken as equal to the length of the probe l and it is infinity for the spherical probe (some authors prefer to use $\pi l/4$ as the upper limit for the cylindrical probe which will change a bit some formulas below [5,6]).

Equations (14) show that knowledge of the profile of the potential near the probe is necessary for calculating the EDF. However, if we assume that $\nu_{en}/v^3 = const$, as it is valid in the approximation for the electron diffusion scattering cross-section in argon, the profile of the potential is not important for calculations. With somewhat less accuracy, the same is valid for other Ramsauer gases (krypton and xenon). Then, the current to the probe of an arbitrary shape is described by expression

$$I_e(V) = \frac{2\pi N_e e S}{m^2}\int_{eV}^{\infty} \frac{f(\varepsilon)d\varepsilon}{\delta(\varepsilon)}, \tag{15}$$

Equation (15) is also valid for any gas in the case of the thin near-probe sheath with thickness much smaller than the probe size (radius). This may be the case for the large electron density.

For the spherical geometry from equations (14) for the EDF we obtain $f_0(w,r) = f_0(w)(1 - a/r)$ and for cylindrical geometry we have $f_0(w,r) = f_0(w)\ln[r/a]/\ln[l/a]$. Since the directed part of the distribution function is $\vec{f_1} = \frac{-v}{\nu_{en}(v)}\nabla f_0$ [13], for the dependence of the electron current on the cylindrical and spherical probes we have

$$I_{ec}(V) = \frac{8\pi^2 N_e l e}{3m\ln[l/a]}\int_{eV}^{\infty} f(w)\frac{v^3}{\nu_{en}}d\varepsilon,\; I_{es}(V) = \frac{16\pi^2 N_e a e}{3m}\int_{eV}^{\infty} f(\varepsilon)\frac{v^3}{\nu_{en}}d\varepsilon, \tag{16}$$

respectively.

If we take into account that factor $\frac{v^3}{\nu_{en}}$ under the integral in equations (16) is equal to $2\lambda(\varepsilon)\varepsilon/m$, then for currents on the spherical and cylindrical probes from equations (16) we obtain expressions

$$I_{ec}(V) = \frac{16\pi^2 N_e l e}{3m^2\ln[l/a]}\int_{eV}^{\infty} f(\varepsilon)\lambda(\varepsilon)\varepsilon d\varepsilon,\; I_{es}(V) = \frac{32\pi^2 N_e a e}{3m^2}\int_{eV}^{\infty} f(\varepsilon)\lambda(\varepsilon)\varepsilon d\varepsilon, \tag{17}$$

respectively. It is clear that equations (17) can be obtained also in the sink theory from equation (9). Consequently, for the Ramsauer gases or thin near-probe sheath in the diffusion regime ($a \gg \lambda$), the sink theory and the kinetic analysis give the same expressions (17) for the currents to probes. It is also possible to get expressions for the EDF similar to equation (10)

$$f(\varepsilon) = \frac{-3m^2 \ln[l/a]}{16\pi^2 N_e l e^3 V \lambda} \frac{dI_{ec}}{dV} \quad \text{and} \quad f(\varepsilon) = \frac{-3m^2}{32\pi^2 N_e a e^3 V \lambda} \frac{dI_{es}}{dV} \tag{18}$$

for cylindrical and spherical probes, respectively.

3.2. Arbitrary case ($\lambda_\varepsilon \gg a$)
Reduction of the gas pressure leads to increasing λ_e and will make the condition $\overline{a} \gg \lambda_e$ invalid. The kinetic approach may also provide a solution for this case. In this case the boundary condition $f(\varepsilon \geq eV, a) = 0$ is not satisfactory [14-16]. It leads to the discontinuity of the distribution function at the point $w = eV$. This in turn can lead to the following paradox [14]: under certain conditions, taking into account the sink effect, the electron current to the probe, calculated according to equation (3), can be greater (!), than the current to the probe in a collisionless-transport mode. This is due to the fact, that in some situations the value of $\delta(\varepsilon)$ is not large enough compared to unity for a given type of the EDF (although the sink parameter can reach values of between 30 and 50). Therefore, it is necessary to be careful to use expressions (15) for the EDF which rapidly falling with energy, especially in the region of large retarding potentials. The use of the original expression (3) maybe more preferable.

If we take

$$f_0(w \geq eV, a) = \frac{4}{3} f_1(w \geq eV, a),$$

as the second boundary condition, then the electron current to the probe will be determined by expression

$$I_e(V) = \frac{2\pi N_e e S}{m^2} \int_{eV}^{\infty} \frac{f(w)(w - eV)dw}{1 + \frac{3}{4} a^\alpha (w - eV) I(w, a)}, \quad I(w, a) = \int_r^A \frac{dr}{r^\alpha \lambda(\varepsilon)\varepsilon}, \tag{19}$$

where $A = \infty$ for a spherical probe and $A = l$ for a cylindrical probe. It is seen, that solution (19) for Ramsauer gases, when $\lambda(\varepsilon)\varepsilon \approx$ const and the potential profile is insignificant, practically coincides with equation (3) in the diffusion regime.

Once again, in the general case, knowledge of the potential profile is required to find the dependence of the electron current on the probe on the retarding potential. In this case, probe plasma diagnostics become very difficult. On the other hand, one can try to numerically estimate the value of the integral

$$I(w, a) = \int_r^{a+d} \frac{dr}{r^\alpha \lambda(\varepsilon)\varepsilon} + \frac{1}{\lambda(w)w} \int_{a+d}^{A} \frac{dr}{r^\alpha}, \tag{20}$$

(outside the sheath/pre-sheath we assume that $\varphi = 0$ and $\varepsilon = w$). Clearly valid, for the thin sheath (or small retarding potentials) in the diffusion-transport mode ($a \gg \lambda$), is the integral $I(w, a) \cong ((\lambda(w)w)^{-1}) \times \ln[l/a]$. Likewise valid, for the current to the probe obtained above in the framework of the theory of the sink effect, are expression (3) and equation (19), obtained in the kinetic approach. Here we note that the choice of such an approximation for $I(w, a)$ and for nonzero V turns out to be sufficiently accurate [4], as estimates show.

For the case of the collisionless-transport mode ($a \ll \lambda$), when condition $\overline{a} \ll \lambda_\varepsilon$ is automatically satisfied, equation (19) implies equation (2) obtained by Druyvesteyn under the condition of absence of collisions between atoms and electrons flowing to the electric probe. The measurements of the EDF is also possible when collisions in the sheath/pre-sheath are taken into account and instead of condition $a \ll \lambda$, it is sufficient that a much weaker requirement $\overline{a} \ll \lambda_\varepsilon$ is satisfied. This circumstance, first noted in Ref. [14], has a simple physical meaning: electrons, which come to the probe from the unperturbed plasma, lose an insignificant fraction of their energy even in collisions in the sheath/pre-sheath.

So, if we ignore the influence of the near-probe sheath, the theory of the sink effect and the kinetic theory give virtually identical descriptions of the current to the probe in the diffusion regime. Obviously, for the zero-retarding potential, the saturation currents completely coincide if, in the theory [8], we assume that the motion of electrons to the probe is diffusive up to its surface. The importance of the kinetic approach [12] lies in the fact that it allowed us to drop the condition $a << \lambda$ and estimate the influence of collisions in the sheath. The main point is that the kinetic approach helped to indicate the limit ($\overline{a} \ll \lambda_\varepsilon$) of the parameters in the plasma, for which measurements of the EDF are possible.

We note that expression (19), for the case of a spherical probe, was first obtained in Ref. [17]. In Ref. [17], also for the first time, to the best of our knowledge, the Monte Carlo method was used for simulating the collection of electrons by such a probe for a thin sheath of space charge. As the simulation for $\lambda = const$ shows, while a/λ increments across the range from 3 to 10, the electron current can be described within the limits of error bar as

$$I(V) = \frac{2\pi N_e eS}{m^2}\int_{eV}^{\infty}\frac{f(w)(w-eV)dw}{1.3+\delta(1-\frac{eV}{\varepsilon})},\quad \delta = \frac{3a^2}{4\lambda(a+\lambda)}. \tag{21}$$

An attempt was made to refine the calculation of currents for cylindrical and spherical probes in Refs. [15,16] with a boundary condition that is slightly different from that used in Refs. [17]. The fact is that, in these works, the boundary condition $f_0(w \geq eV, a) = \frac{4}{3}f_1(W \geq eV, a)$ was used, which, strictly speaking, is valid only for completely isotropic EDF. Near the surface of the probe, the EDF associated with the case of small V is noticeably anisotropic and, in Ref. [15,16], the so-called effective boundary condition $f_0(W \geq eV, a) = \gamma(\frac{a}{\lambda})f_1(W \geq eV, a)$ [4,18] was accounted for in the EDF calculation.

The results for the dependence of the currents on the retarding potential for such a choice can be presented in the form

$$I(V) = \frac{2\pi N_e eS}{m^2}\int_{eV}^{\infty}\frac{f(w)(w-eV)dw}{\frac{3}{4}\gamma+\frac{3}{4}a^{\alpha}(w-eV)I(w,a)}. \tag{22}$$

The coefficient $\gamma = \gamma(\frac{a}{\lambda})$ in equation (22) varies slightly from 4/3 to 0.71 as $\frac{a}{\lambda}$ changes from zero to infinity. In the case of a spherical probe, for which $x = \frac{a}{\lambda} > 1$, we have $\gamma = 0.71 + 0.4/x$.

So, to date, a number of slightly different descriptions of the current to the probe have been proposed, using different boundary conditions. Obviously, in the diffusion limit and for a thin sheath, interpreting the current-voltage characteristic using different boundary conditions should give practically identical results. Further, it should be expected that, at zero retarding potential, effective boundary conditions will yield a closer result to the simulation results (if they are accurate enough). Assuming that the model current (17) is found to an accuracy of 1-2%, we estimate the accuracy of the use of equations (3), (19), (22) to be comparable.

Below, the electron current (3) obtained by Swift [8] is denoted by $I_s(V)$; current (19) with the unity in puncture the integrand as I_{10} (V) (index 0 should remind us that the sheath is infinitely thin), similarly the current (22) is like $I_{\gamma 0}(V)$, simulated by the Monte Carlo current method (21) as $I_m(V)$. The relationship of the currents (3), (19), (22) to the simulated one is shown in figure 1(a, b, c) for the values $a/\lambda = 3$ and $a/\lambda = 10$, and $a/\lambda = 50$. As can be seen, the effective boundary conditions give more accurate values only for the saturation currents. However, as the retarding potential V (given in figure 1 in units of electron temperature) increases, the discrepancy between the currents $I_{\gamma 0}$ and I_m (V) increases noticeably. For example, for $V = 6$, the discrepancy reaches 40% even at $a/\lambda = 10$ while, for $a/\lambda = 3$, the discrepancy exceeds 50%. On the other hand, for $a/\lambda = 10$, the currents $I_s(0)$ and $I_{10}(0)$ differ by only about 4% from $I_m(0)$. Consequently, the accuracy, for example, I_{10} (V) for large V is not worse than 8% (interestingly, this is only 7% greater than the accuracy of I_s (V)). For large $a/\lambda = 50$, the differences in the currents (3), (19), (22) do not exceed 10-15%, which we interpret as the precision of the prediction.

Thus, current (19) is better than current (22) in terms of corresponding to the model current. This is a consequence of the fact that the effective boundary conditions are valid only for neutral particles. Simplifying (without taking into account the influence of the sheath) the description of the current within the framework of the sink theory (3) also gives a very good approximation. It is easy to see that if we discard the influence of the sheath and assume the motion of the electrons to the probe diffusively down to its surface, for the sink parameter, we have

$$\delta(\varepsilon) = \frac{3a}{4\lambda(\varepsilon)},$$

which exactly coincides with the sink parameter in the thin sheath limit.

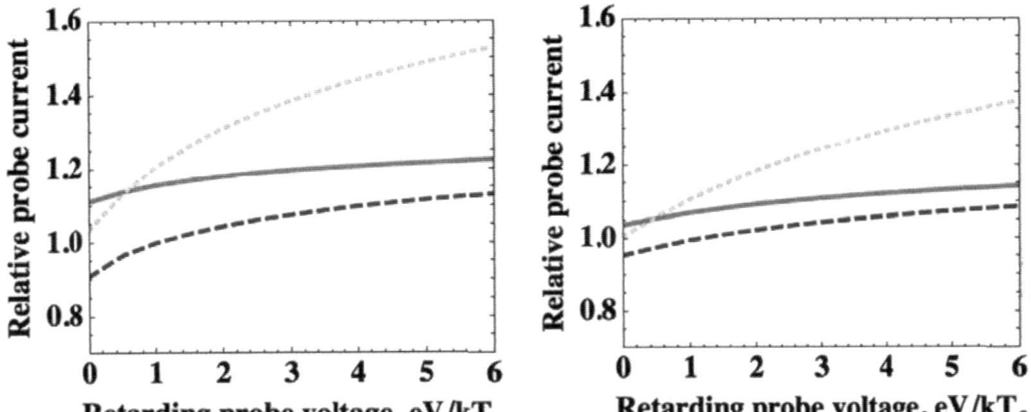

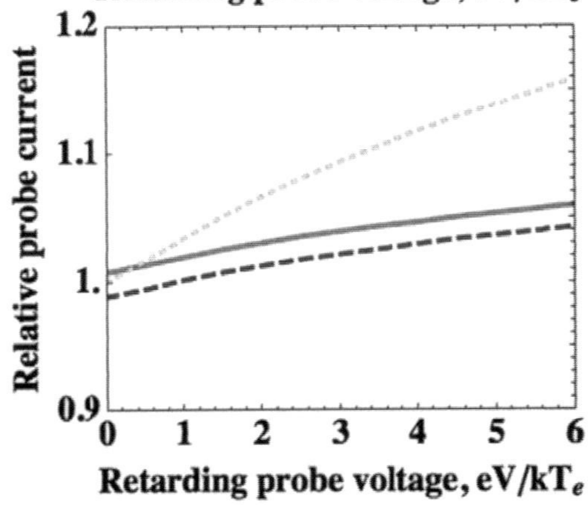

Figure 1. The relationship between the electron currents of a spherical probe for a thin sheath of space charge: (top/left) a/λ = 3, (top/right) a/λ = 10 and (bottom) a/λ = 50. I_s/I_m (solid), I_{10}/I_m (dashed) and $I_{\gamma 0}/I_m$ (dotted).

4. Experimental measurements at high gas pressures

Starting with the analysis, we first note that experimental measurements of the EDF are practically absent: only a very few papers are known to the authors. Let's start with the experiment [12], where electro-kinetic characteristics were measured in a discharge plasma in argon at pressures up to 100 Torr. In that work, it was concluded that it is possible to measure the EDF and the electron density according to equations (17). As shown in [11], this conclusion, generally speaking, is not confirmed by the experimental data given in the same work [12].

We recall that the measurements were made in a positive column of the argon discharge using a cylindrical probe of radius a = 0.1 mm and a length l = 0.3 cm located on the axis of the discharge tube. The probe current-voltage characteristics provided in Ref. [12] show that, for example, at a pressure p = 62 Torr and for the discharge current of 35 mA, the electron saturation current is approximately 0.45 mA. Let us estimate the value of the sink parameter for the experimental conditions of Ref. [12] and compare it with the calculated value. For the discharge current, I_d through the tube and for the electron saturation current to the probe, I_e^{sat}, we have, respectively,

$$I_d = 0.43N_e(0)eS_d v_d \text{ and } I_e^{sat} = \frac{N_e(0)eS\bar{v}}{4(1+\delta)}. \tag{23}$$

where $N_e(0)$ is the electron density at the center of the discharge tube, S_d is the area of its cross section, and v_d is the drift velocity of the electrons. Thus, in order to estimate the value of the electron discharge parameter for the probe, we obtain the relation

$$\delta \approx \frac{I_d S\bar{v}}{I_e^{sat} S_d 1.7 v_d} - 1.$$

Under the experimental conditions, the average E/N value was $3.5 \cdot 10^{-17}$ Vcm^2 and, according to [19,20], the drift velocity is $4.5 \cdot 10^5$ cm/s. The characteristic electron energy is from 5 to 7 eV, so we assume the value of the average velocity to be $\bar{v} = 10^8$ cm/s. Thus, for an experimentally determined value of the sink parameter and taking into account $S/S_d = 6.6 \cdot 10^{-4}$, we obtain $\delta_{exp} = 5$. At the same time, the calculated value of the sink parameter in the case under consideration is 50! That is, an order of magnitude difference is observed. This, of course, is a rough estimate since, in the experiment [12], the EDF is far from Maxwellian and has the form $f(\varepsilon) \sim \exp(-(\varepsilon/\bar{\varepsilon})^4)$ with a characteristic energy from 6 to 7 eV. More accurate calculations, when the saturation current is calculated according to equations (16) with the experimentally found EDF (with the electron density estimated by the current of the tube), show the discrepancy between theory and experiment to be a factor of approximately from 3 to 3.5.

The following experiments [11] were carried out in helium gas discharge plasma using the current-voltage characteristics of the "thick" (a_1 = 0.6 mm) and "thin" (a_2 = 0.045 mm) cylindrical probes at pressures from 0.9 to 3.3 Torr. In the chosen pressure range, the thicker probe sink parameter reached values of $\delta \approx 10$, while the thin probe parameter δ is less than 1. Since $N_e \sim I_e^{sat}(0)(1+\delta)$, the ratio of the saturation currents of the two probes, under the same experimental conditions, will be determined by

$$\frac{I_{e1}^{sat}(0)}{I_{e2}^{sat}(0)} = \frac{S_1}{S_2}\frac{1+\delta_2}{1+\delta_1},$$

where $I_{e1}^{sat}(0)$, $I_{e2}^{sat}(0)$, S_1, S_2, δ_1 and δ_2 are saturation currents, areas and sink parameters for probe 1 and probe 2 respectively. Thus, a comparison of the experimentally obtained and calculated ratios of saturation currents makes it possible to judge the applicability of the sink theory for finding the plasma parameters. Such a comparison of the ratios shows that, at the maximum pressure (3.3Top), there is a noticeable (approximately threefold) increase in the experimental ratios of the saturation currents over the calculated ones, that is similar to obtained in [12].

What is the reason for the observed discrepancy between theory and experiment? Here we can note a number of factors. The main factor is, apparently, the presence in the gas-discharge plasma of the electric field, which causes the transfer of electrons from the unperturbed plasma to the depleted region of the probe region. Since in the theory it is assumed that the current to the probe is determined only by the gradient drift, for its applicability it is necessary to have the ratio

$$\xi = \frac{{}^{eE}/_{mv}\,{}^{\partial f}/_{\partial v}}{{}^{\partial f}/_{\partial r}} \approx \frac{eE\bar{a}}{\bar{\varepsilon}} \ll 1, \tag{24}$$

where $\bar{\varepsilon}$ is the characteristic energy of the electron distribution.

Requirement (24) for the case of spherical geometry can be obtained [21] from the following consideration. Solving the equation $\Delta N_e(r) = 0$ for the density profile of $N_e(r)$ under the boundary condition $N_e(a) = 0$ in the diffusion mode yields $N_e(r) = N_e(1 - a/r)$. Hence, for the electron current we have

$$I_e^{sat} = \int D_e \nabla N_e dS = \frac{eSN_eD_e}{a}, \tag{25}$$

(here $D_e = \lambda\bar{v}/3$) which actually coincides with the value of the saturation current

$$I_e^{sat} = \frac{N_e eS\bar{v}}{4(1+\delta_c)},$$

where δ_c=$3a/(4\lambda)$, given by the theory of flow in the diffusion regime.

As shown by (25), the value of the electron density gradient is ~ N_e/a, that is, the spherical probe perturbs the plasma at a distance of ~ a. It is obvious that the electron current transported by an externally applied (created not by a probe) electric field E in a section equal to $\pi(a+a)^2$, should be much less than the diffusive-motion current per probe, that is,

$$\frac{eSN_eD_e}{a} \gg e\mu EN_e 4\pi a^2, \tag{26}$$

or, taking into account the relation $\frac{\mu_e}{D_e} = e/(kT_e)$, we find the condition $\frac{eEa}{kT_e} \ll 1$.

Calculation [20] for a spherical probe shows that the saturation current for a Maxwellian EDF with allowance for the external electric field increases approximately in proportion to the growth of the parameter $\alpha a = eEa/(2kT_e)$. Unfortunately, such calculations for a cylindrical probe are not known to us.

When speaking about estimates of the applicability limits (while satisfying the inequalities $\bar{a} = a + d \ll \lambda_\varepsilon$ and (28)) of the theory for a cylindrical probe, we note that there are significantly different definitions of the quantity $\bar{a}$. For example, in Ref. [12], we can say, the upper, or conservative, limit is $\bar{a} = l/2$. Probably a more accurate (mid-conservative) estimate can be obtained by setting the unperturbed EDF at a distance where $N_e(\bar{a}) = 0.5N_e$. Then, for a cylindrical probe, we have $\bar{a} \approx a\sqrt{l/a}$ (with an analogous choice for a spherical probe $\bar{a} = 2a$). Finally, the lower estimate can be obtained by analogy with a spherical probe. It is $\bar{a} = a\ln[l/a]$ [22]. With this choice of $\bar{a}$, by analogy with a spherical probe (that is, analogously to equation (26)) for a probe perpendicular to the field (for large electric fields it is advisable to place a probe perpendicular to electric field E), one can obtain

$$\frac{eSN_eD_e}{a\ln[l/a]} \gg e\mu EN_e 2al(1+\ln[l/a]), \tag{27}$$

whence we find the requirement

$$\xi = \frac{eEa\ln[l/a])(1+\ln[l/a])}{kT_e\pi} = \frac{eE\bar{a}}{kT_e}\frac{(1+\ln[l/a])}{\pi} \ll 1. \tag{28}$$

Further, the usual estimate $\bar{a} \ll \lambda_\varepsilon$ was given in Ref. [12] within an order of magnitude. In fact, this condition looks more exactly like $(\lambda_\varepsilon/\bar{a})^2 \gg 1$ [21]. Then, when choosing the lower limit, it is likely that the condition $\lambda_\varepsilon > 3\bar{a} = 3a\ln[l/a]$ will be sufficient for the feasibility of the theory of the sink effect, as in this case $(\lambda_\varepsilon/\bar{a})^2 \sim 10$.

As shown by the experimental data [11,12], the conditions (24) (with the choice of $\bar{a} = l/2$) or (28) (where $\bar{a} = a\ln[l/a]$) are certainly not satisfied. For example, when choosing $\bar{a} = l/2$ and the maximum test pressure in [11], the value ξ was of the order of unity and, in experiments in Ref. [12], it was from 2 to 2.5. Even when choosing the lower estimate for $\bar{a} = a\ln(l/a)$, we have $\xi \approx 1$ in Ref. [12] and $\xi \approx 0.35$ in Ref. [11].

Let us now briefly discuss the possible influence of ionization processes in the probe sheath on the probe currents. Expression for the sink parameter $\delta \approx \frac{3a}{4\lambda}\ln[l/a]$ shows that when the probe length l tends to infinity, the current density vanishes. At the same time, obviously, when recombination and ionization of particles in the near-probe sheath in the plasma are taken into account, the current density j_e on such a probe must be finite [23]. It was shown in Ref. [24] that for an infinite probe at the center of a discharge tube of radius R, the electron density

$$j_e \approx [1 + \frac{3a}{4\lambda}\ln[R/(2.4a)]^{-1}, \tag{29}$$

i.e., instead of expression $\delta = \frac{3a}{4\lambda}\ln[l/a]$, when ionization in the plasma volume is taken into account, we have $\delta = \frac{3a}{4\lambda}\ln[R/(2.4a)]$. It is clear that in the case when the ionization length is

$$L_i = \sqrt{{D_a}/{z_i}} = {R}/{2.4} \leq l,$$

the known expression for the sink parameter, generally speaking, can no longer be used. Interestingly, in most studies of a gas-discharge plasma, tubes of a radius from 1 to 2 cm are used, the characteristic probe length is from 0.5 to 1 cm, i.e., l and L_i are practically equal.

Hardly, however, this circumstance can strongly affect the value of the saturation current, because one can expect that the "effective" sink parameter is described by an expression of the type of

$$\delta_{eff} \approx \frac{3a}{4\lambda}\ln\left[\frac{L_i l}{(l + L_i)a}\right],$$

i.e. it is determined by the geometric mean of L_i and l. If we assume that δ_{eff} is determined by this relation, then under the experimental conditions [12], the decreasing δ_{eff} in comparison with δ is only 5%. Consequently, one does not expect such a significant divergence between theory and experiment. At the same time in Ref. [12], a decrease of δ_{eff} in comparison with δ can be about 40%. On the other hand, the role of ionization waves (traveling striations) is completely unclear, when, according to [25], the densities of charged particles at the maximum and minimum of the wave can differ by several times.

In conclusion, let us briefly consider the possible error in finding the sink parameter. As noted in Ref. [11], the sink theory for the diffusion regime for $\lambda = const$ and the Maxwellian EDF gives the same values of the saturation current as the Bohm theory [2]. For example, for a cylindrical probe represented by an ellipsoid of revolution, we have

$$I_e^{sat} = \frac{N_e e S \vec{v}}{4(1+\delta_E)}, \text{ where } \delta_E = \frac{\bar{v} S}{16\pi C_S D_e}, \tag{30}$$

where C_S and D_e are the capacity of the body in the form of an ellipsoid with semi-axes a, $l/2$ and the diffusion coefficient of electrons $D_e = {\lambda\bar{v}}/{3}$. The capacity C_S is then calculated according to

$$C_{se} = \frac{l\sqrt{1-(2a/l)^2}}{2\tan^{-1}\sqrt{1-(2a/l)^2}}, \tag{31}$$

where subscript "se" is for an elliptical body.

At the same time, there are calculations (see, for example, [26]), which for a cylindrical segment (subscript "sc") of length l and radius a give (here $\Lambda = \ln(l/a)$)

$$C_{sc} = \frac{l}{2\ln[l/a]}\left(1 + \frac{1-\ln 2}{\Lambda} + \frac{1+(1-\ln 2)^2 - \frac{\pi^2}{12}}{\Lambda^2} + O(\Lambda^{-3})\right). \tag{32}$$

The ratio $\psi(x) = C_{sc}/C_{se}$, as a function of $x = l/a$, is shown in figure 2. As can be seen, the capacity of the cylinder is always greater than the capacity of the ellipsoid. For a prolate probe with values l/a in the range from 50 to 250, generally speaking, the sink parameter should be reduced by an average of approximately 5-8%.

It is interesting that for $l/a \sim 10$ such a change can already be from 20% to 25%. In this connection, we note that the decrease in the experimental δ_{exp} in comparison with the calculated δ for a thick probe observed in the experiment [12] is apparently partly due to this factor. For Ref. [12], this decrease is about 12%.

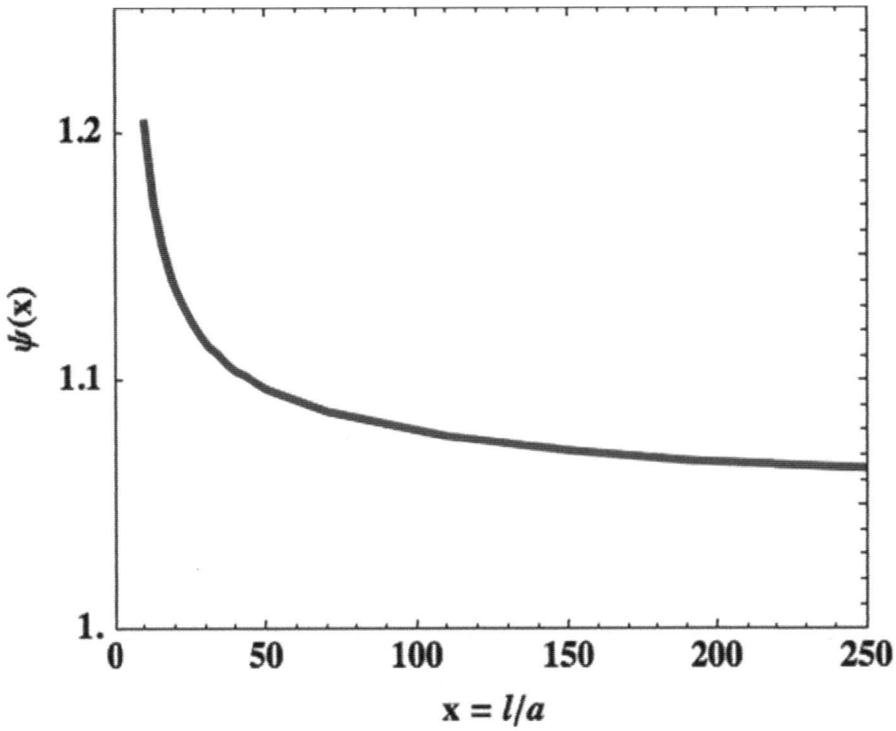

Figure 2. Function ψ depending on parameter x = l/a.

5. Large wall probes

The methods of probe measurements in plasmas at high pressure, described above, are based on the assumptions of the EDF nonlocality in a plasma and that the probe disturbs the plasma properties within a small fraction of electron energy relaxation length λ_ε only. In Ref. [27], an application of a *small* wall probe for the EDF measurements, in the case when the plasma dimension is smaller than λ_ε, has been discussed. Typically, a wall probe is an electrically isolated segment of the plasma volume wall which collects the current from the plasma for different probe potentials. An area of a small probe is much smaller that plasma wall area. The wall probe does not require a probe holder and thus may reduce the disturbance of plasma. The distortion of the EDF measurements by the ion current may be significantly reduced due to the much greater probe radius of curvature than for a cylindrical or spherical probe. It is probably possible to modify the above probe theories to conduct absolute measurements with small wall probe [28]. However, small wall probes typically cannot measure the EDF for energies below from 3 to 5 average electron energies in the central part of the plasma which has lowest electric potential. Those electrons are imprisoned in the volume and cannot reach plasma walls.

There are a number of publications about the EDF measurements with such probes (see Ref. [29] and references therein). However, they are inconclusive about real possibility and reliability of measurements with small wall probe. At present the small wall probes cannot be recommended for reliable practical use and require additional confirmation and validation.

A method for electron energy spectra studies using a *large* wall probe has been developed in Ref. [30]. The area of the large wall probe may be significantly larger than for a typical spherical or cylindrical Langmuir and small wall probe resulting in a drastic increase in the probe sensitivity. That is especially important for the measurements of energetic electrons in the tail of the EDF as their density is much smaller than the density of thermal electrons. It was demonstrated that the large wall probe allows us to conduct measurements for gas pressures from a fraction of one Torr to atmospheric pressures [6,30,31] and maybe higher. The theory of the large wall probe has not been developed yet. However, it was experimentally demonstrated in different gases at pressures of about one Torr, that, in the used experimental device, the EDF is proportional to the *second* derivative of electron probe current I_e with respect to the probe potential V [30].

The sketch of the experimental device used in Ref. [30] is shown in figure 3. The discharge takes place between a plane disc-shaped molybdenum cathode (C) and anode (A), schematically shown in figure 3. The plasma volume is bounded by a cylindrical stainless-steel wall (W). The cathode and anode are 2.5 cm in diameter. The distance between the cathode and anode is 1.2 cm. The wall W was used as a large wall probe.

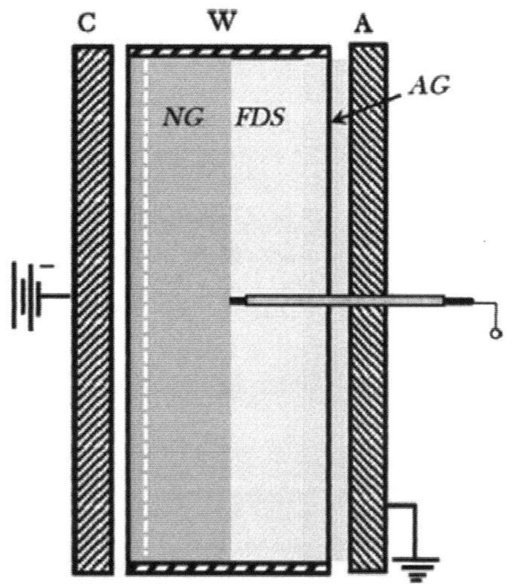

Figure 3. Schematic diagram of experimental device: cathode C, grounded anode A, and cylindrical wall W. Typical structures of the discharge include negative glow (NG), anode glow (AG) and Faraday dark space FDS. The cathode sheath boundary (located in the NG) is indicted by the dashed line.

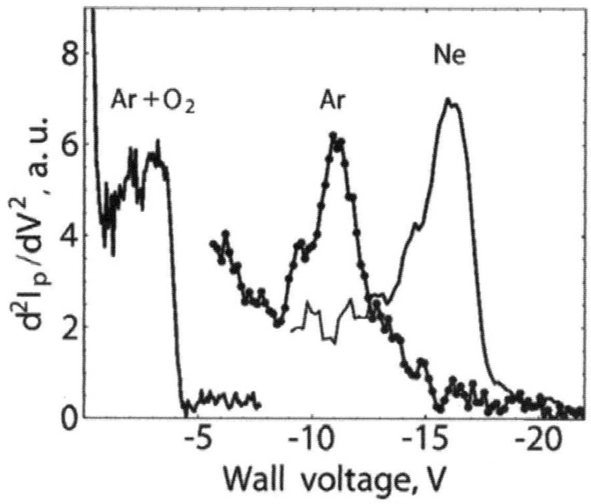

Figure 4. High-energy portion of d^2I/dV_2 (absolute value) in neon (3 Torr), argon (0.5 Torr) and oxygen (20%)/argon (80%) (0.5 Torr) dc discharge. The discharge currents were 10 mA, 2mA and 3 mA, respectively. The maxima at 16 eV and 11.5 eV are due to collisions of neon and argon metastable atoms with slow electrons. The maximum at $\approx$ 4 eV is due to electron detachment from oxygen, i.e. $O^- + O \rightarrow O_2 + e$ [30].

The plasma of the device typically contains two group of electrons. Low-energy groups of thermal electrons with Maxwellian distributions have electron temperatures at the order of 0.1 eV. The distribution of energetic electron groups is far from equilibrium and exhibits several peaks in the EDF tail. The density of these electrons maybe from 10^4 to 10^6 times less than that of the low-energy electrons. The origination of the high-energy electrons is collisions of metastable excited atoms between themselves and with electrons in the reactions

$$\begin{aligned} A^* + A^* &\rightarrow A^+ + A + e, \\ A^* + A^* &\rightarrow A_2^+ + e, \\ A^* + e &\rightarrow A + e. \end{aligned} \tag{33}$$

Here, A^* denotes an exited atom, and A^+ and A_2^+ denote atomic and molecular ions. The first two reactions in (33) represent the Penning ionization process, whereas the third describes the super-elastic collision of an exited atom with a low-energy thermal electron.

Figure 4 shows the maxima in d^2I/dV^2 measured with the wall probe in the discharge plasma in neon, argon and oxygen (20%)/argon (80%) mixture. The peaks for pure neon and argon at 16.6 eV and 11.5 eV are the result of collisions of slow thermal electrons with the excited atoms (the third reaction in (33). The small peak in the oxygen argon mixture near 4 eV is due to electron detachment from negative oxygen ions. The energetic electrons from the detachment have energy of 3.6 eV.

Figure 5 shows d^2I/dV^2 in the afterglow of a helium/argon mixture dc pulsed discharge. During the discharge, the plasma contains energetic primary electrons accelerated in the near-cathode sheath in the direction of the anode. These electrons ionize the gas thus producing the plasma. They also create metastable atoms, which can in turn generate groups of energetic electrons. Upper curve is 85 μs after the pulse and lower curve is 450 μs. The maximum at about 4 eV is connected to Penning ionization of argon atoms by helium metastable. The maxima at about 14.5 eV and about 20 eV are deactivation of helium metastable atoms by slow electrons.

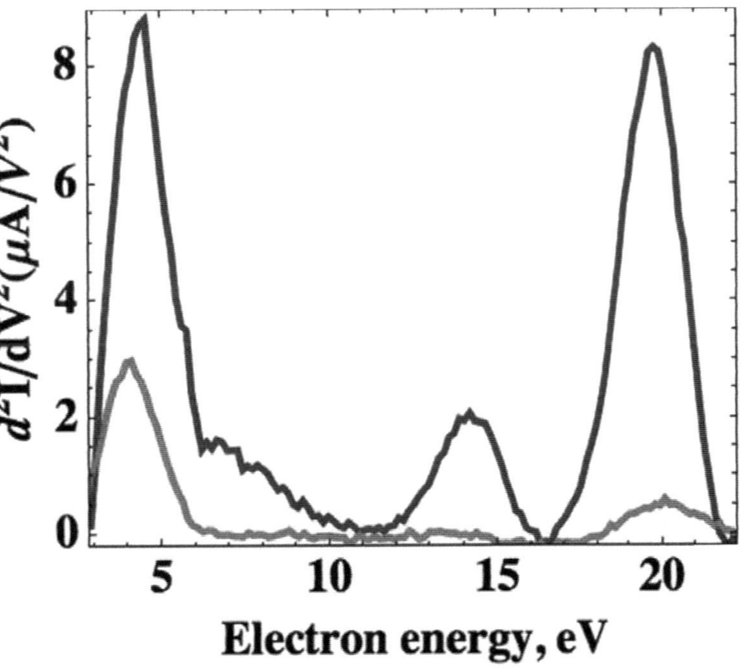

Figure 5. d^2I/dV^2 in the afterglow of a helium/argon mixture (4 Torr, argon is 0.002%) dc pulsed discharge. Upper (blue) curve is 85 μs after the pulse and lower (red) curve is 450 μs). The maximum at 4 eV is connected to Penning ionization of argon atoms by helium metastable. The maxima at 14.5 eV and 20 eV are deactivation of helium metastable atoms by slow electrons.

The area under a peak in the energy spectrum is proportional to the density of energetic electrons resulting from a specific reaction. The ratio of different constituents of a gas mixture can be obtained from the areas under the peaks corresponding to those constituent reactions. For example, the ratio of the peak areas at 4, 15 and 20 eV in figure 5 yields the ratio between argon density atoms, helium metastable atoms and slow electrons. Furthermore, the area of the maximum at 4 eV declines with time in afterglow due to reduction of helium metastable atom density as argon atom density is unchanged during afterglow. As a calibrated gas mixture has been used with 0.002% of argon (density of argon atoms N_{Ar} is 2.8×10^{12} cm^{-3}), density of metastable helium atoms and electrons can be measured. It was found for the blue curve that density of metastable helium atom N_{He} is 2.2×10^{11} cm^{-3} and density of electrons N_e is 3.6×10^{11} cm^{-3}. For red curve $N_{He} = 6.2 \times 10^{10}$ cm^{-3} and $N_e = 1.6 \times 10^{11}$ cm^{-3}.

6. Conclusions

Thus, at present a number of theories exist, which allow probe measurements of the EDF under high gas pressures (up to an atmospheric pressure and higher) for which the simple Druyvesteyn formula cannot be used. So far, those theories have been used in a very limited number of works, which does not allow a reliable check of their accuracy. Those theories (especially for the arbitrary case) probably cannot provide accurate interpretations of measurements for high pressure comparable with the Druyvesteyn formula (for low pressure), as they are more complicated and take into account many additional and sometime unknown factors (like plasma potential behavior near the probe). Analysis of a few experimental data shows that, due to a number of reasons, the use of high pressure theory for the processing of probe *IV* traces may lead to significant measurement errors. For high pressure, influence of a probe-holder may be greater than for lower pressure. Electron reflection from dirty probe surface may also be more important for higher pressure [32]. Other effects, which can distort the probe measurements enumerated, for example, in reviews [4,5], maybe also more significant for higher pressure EDF measurements. However, for a definitive answer to the question of how these factors affect the results of the experiment, of course, additional studies are required. A small wall probe can be a useful for some measurements but, so far, no reliable experimental confirmation exists. A large wall probe can provide higher sensitivity, but results may be spoiled by variation of plasma properties under changed probe potential [33,34]. Nevertheless, the large wall probe may be useful, for example, in analysis of energetic electrons in plasmas.

References

[1] Mott-Smith H M and Langmuir I 1926 *Phys. Rev.* **28** 727-63
[2] Kagan Y M and Perel V I 1964 *Sov. Phys.—Usp.* **6** 767
[3] Swift J D and Schwar M J R 1970 *Electrical Probes for Plasma Diagnostics* (London: Iliffe Books)
[4] Demidov V I, Kolokolov N B and Kudryavtsev A A 1996 *Probe Methods for Low-Temperature Plasma Investigation* (Moscow: Energoatomizdat)
[5] Demidov V I, Ratynskaia S V and Rypdal K 2002 *Rev. Sci. Instrum.* **73** 3409
[6] Godyak V A and Demidov V I 2011 *J. Phys.* D: Appl. Phys. **44** 233001
[7] Druyvesteyn M J 1930 *Z. Phys.* **64** 781-98
[8] Swift J D 1962 *Proc. Phys. Soc.* **79** 697-701
[9] Lukovnikov A I and Novgorodov M Z 1971 *Brief Communications Physics* #1, 27
[10] Devyatov A M and Malkov M A 1984 *Sov. Phys. J.* **27** 197
[11] Malkov M A 1991 *High Temp.* **29** 329
[12] Golubovsky Y B, Zakharova V M, Pasunkin V I and Tsendin L D 1981 *Sov. J. Plasma Phys.* **7** 340
[13] Mitchner M and Kruger C H 1973 *Partially Ionized Gases* (New York: Wiley) 518 p
[14] Malkov M A 1985 *Probe Diagnostics with Electron Sink Effect Taken into Account and for Anisotropic EDF* (Moscow: Moscow State University)

IOP Conf. Series: Journal of Physics: Conf. Series **958** (2018) 012003 doi:10.1088/1742-6596/958/1/012003

[15] Gorbunov N A, Kolokolov N B and Kudryavtsev A A 1989 *Sov. J. Plasma Phys.* **15** 1513-19
[16] Arslanbekov R R, Khromov N A and Kudryavtsev A A 1994 *Plasma Sources Sci. Technol.* **3** 528
[17] Friedland L and Kagan Yu M 1979 *J. Phys.* D: Appl. Phys. **12** 739-42
[18] Davison B 1958 *Neutron Transport Theory* (Oxford: Oxford University)
[19] Brown S 1959 *Elementary Processes in Gas Discharge Plasma* (Cambridge: MIT)
[20] Nakamura Y and Kurashi M. 1988 *J. Phys.* D: Appl. Phys. **22** 718-25.
[21] Devyatov A M, Kuzovnikov A A and Malkov M A 1991 *Vestnik Moscow State University* #1 26-30
[22] Malkov M A 2014 *Plasma diagnostics with Langmuir probe* (Saransk: SSU) 247 p
[23] Chen F F 1965 *Plasma Diagnostic Techniques*, ed R H Huddlestone and S L Leonard (New York: Academic) p 113
[24] Devyatov A M and Malkov M A 1984 *Vestnik Moscow State University* #4 68-70
[25] Pekarek L 1965 *Sov. Phys.-Uspekhi* **94** 463-500
[26] Jackson J D 2000 *Am. J. Phys* **68** 789–99
[27] Kudryavtsev A A 1999 A new method of gas Analysis using Penning-electron energy spectra, *Proc. IWRFRI'99, Russian Technologies for Industrial Applications. May 24-26, 1999, St. Petersburg, Russia, p. 94*
[28] Demidov V I and Kudryavtsev A A 2014 *Phys. Plasmas* **21** 093506
[29] Kudryavtsev A A, Saifutdinov A I, Stefanova M S, Pramatarov P M, and Sysoev S S 2017 *Phys. Plasmas* **24** 054507
[30] Demidov V I, Adams S F, Blessington J, Koepke M E and Williamson J M 2010 *Contrib. Plasma Phys.* **50** 808-13
[31] Demidov V I, Adams S F, Miles J A, Koepke M E and Kurlyandskaya I P 2016 *Phys. Plasmas* **23** 103508
[32] Demidov V I, Adams S F, Kaganovich I D, Koepke M E and Kurlyandskaya I P 2015 *Phys. Plasmas* **22** 104501
[33] Demidov V I, Kudryavtsev A A, Kurlyandskaya I P and Stepanova O M 2014 *Phys. Plasmas* **21** 094501
[34] Adams S F, Demidov V I, Bogdanov E A, Koepke M E, Kudryavtsev A and Kurlyandskaya I P 2016 *Phys. Plasmas* **23** 024501

Kinetic Analysis of Weakly ionized Plasmas in presence of collecting walls

J. Gonzalez, J. M. Donoso

Department of Applied Physics, Escuela Técnica Superior de Ingeniería Aeronáutica y del Espacio, Universidad Politécnica de Madrid, 28040 Madrid, España

E-mail: jorge.gonzalez@upm.es

Abstract. Description of plasmas in contact with a wall able to collecting or emitting charged particles is a research topic of great importance. This situation arises in a great variety of phenomena such as the characterization of plasmas by means of electric probes, in the surface treatment of materials and in the service-life of coatings in electric thrusters. In particular, in this work we devote attention to the dynamics of an argon weakly ionized plasma in the presence of a collecting wall. It is proposed a kinetic model in a 1D1V planar phase-space geometry. The model accounts for the electric field coupled to the system by solving the associated Poisson's equation. To solve numerically the resulting non-linear system of equations, the Propagator Integral Method is used in conjunction with a slabbing method. On each interrelating plasma slab the integral advancing scheme operates in velocity space, in such a way that the all the species dynamics dominating the system evolution are kinetically described.

1. Introduction

The description of plasma-wall interactions is an open problem of major importance in experiential plasma physics since the dynamics of light and heavy species close to a wall bears on probe measurement processes, service life of walls and coatings and surface treatment procedures. This interaction appears in experiments carried out involving plasmas [1, 2, 3] as well as in vessels flying in space [4]. Many works have provided fluid models, as the those by T. Gyergyek *et al.* [5, 6, 7] or kinetic descriptions, as the works by M. D.Campanell *et al.* [8, 9] or J. P.Sheehan *et al.* [10, 1], among many other authors [11, 12, 13, 14, 15, 16, 17]. An important case of plasma-wall interaction appears in the characterization of plasmas, usually carried out with emissive [18, 19, 20] or collecting probes [21, 22, 3] to calculate plasma parameters, such as species temperature, density or the plasma potential, responsible of the energy transference. The description of this interaction is usually established by models that may not capture all microscopic processes involved, specially in Weakly Ionized Plasmas (WIP), for which the inclusion of a dense heavy cold neutral species hinders the development of such models [21, 23, 24]. Our research group is actually interested in the use of collecting probes to analyse the low temperature plasma generated in our laboratory [25, 26, 27], making this topic of main interest in the characterisation of, for example, the plasma plume produced by an electric thruster [25].

In this work, some progresses performed to analyse the evolution of Weakly Ionized Plasmas close to collecting walls are presented. In particular, the collection of electrons and ions coming from a quasi–neutral weakly ionized plasma is studied by analysing the different behaviours of ions and electrons through their time evolving dynamics. A theoretical kinetic model is proposed, giving rise to non-linear drift–diffusion equations. The Propagator Integral Method [13, 28, 29] is used to solve the resulting one-dimensional kinetic problem in the velocity space, meanwhile the spatial dependence is introduced by dividing the plasma in a series of interrelating thin slabs. This slabbing method [30, 31] accounts for the flux of particles as a function of the position.

2. Kinetic model and slabbing method

For a plasma bounded by a wall capable of collecting or emitting charged particles, the species dynamics gives rise to a very complex system, related to the fast charge separation that governs the plasma evolution, at the same time modified by the collection of particles. In this section, we present a kinetic model to describe ions and electron dynamics of a weakly ionized plasma close to a metallic wall, with no previous assumptions about the particle density profiles or electric potential shapes. The kinetic model in the 1D1V planar phase–space employed is mathematically expressed as

$$\frac{\partial f_\alpha}{\partial t} + v\frac{\partial f_\alpha}{\partial x} + \frac{q_\alpha E}{m_\alpha}\frac{\partial f_\alpha}{\partial v} = \left(\frac{\partial f_\alpha}{\partial t}\right)_{\alpha 0} \tag{1}$$

$$\begin{aligned} \frac{\partial^2 \phi}{\partial x^2} &= -\frac{e}{\epsilon_0}(n_i - n_e) \\ \phi(0) &= \phi_w \\ \left.\frac{\partial \phi}{\partial x}\right|_{x\to\infty} &= 0 \end{aligned} \tag{2}$$

where $(\partial f_\alpha/\partial v)_{\alpha 0}$ is a Dougherty collision term for charged particles interactions with neutrals, represented by the drift and diffusion parameters in a Fokker—Planck model approach [32, 33] as

$$D_v^{\gamma 0} = -\nu^{\gamma\gamma'}(v - u_0)\,; D_{vv}^{\gamma 0} = \nu^{\gamma\gamma'}(E_\gamma + E_0 - 2u_\gamma u_0), \tag{3}$$

where $\nu^{\gamma\gamma'}$ is the collision frequency, $f_\alpha = f_\alpha(x, v, t)$ is the distribution function for the species $\alpha = e, i$ and ϕ_w is the potential of the collecting wall, located at $x = 0$. To solve this system with an integral advancing stable scheme (PIM), a slabbing method is employed in order to reduce the computational cost that would imply the use of a propagator working in the whole phase-space. This approach allows a better representation of the kinetic processes in the v direction and reduces the number of points required to represent the x dependence. This slabbing method is appropriate since the dynamics in the x direction is dominated by the convective flux of particles, and slightly affected by the diffusive $x - x$ and $x - v$ processes. This slabbing method consists on the numerical evaluation of the flux of particles $v\partial f_\alpha/\partial x$ to replace its corresponding contribution, in (1), by a non–homogeneous term which represents the particle balance through the slab boundaries. Similar approach has been employed in [31], where this effective source term, replacing the so–called Vlasov term, was used to compute the heat flux for transport coefficients calculations of a two-dimensional plasma in the velocity space. Furthermore, this method is based on the pioneer work by J. P. Matte and J. Virmont [30] also used in fully ionized plasma local transport coefficients characterization.

To analyse the behaviour in each different slab, the flux term in the x direction is evaluated as

$$-v\frac{\partial f_\alpha}{\partial x} \simeq s_{flux}(x, v, t) = \begin{cases} -v\frac{f_\alpha(x+\Delta x,v,t) - f_\alpha(x,v,t)}{\Delta x} & \text{if } v < 0 \\ -v\frac{f_\alpha(x,v,t) - f_\alpha(x-\Delta x,v,t)}{\Delta x} & \text{if } v > 0 \end{cases}, \tag{4}$$

coming from the physical meaning of the divergence of $\mathbf{v}f$ as the net flux of a vector field, across the closed smooth surface value of a vanishingly small volume, divided by this volume, both

	Mass (kg)	Velocity (m/s)	Temperature (eV)	Density (m^{-3})
e	$9.10938356 \times 10^{-31}$	$\sqrt{\frac{kT_e}{m_e}}$	1	$r_i \frac{n_T}{2}$
i	$72819.6 \times m_e$	$\sqrt{\frac{kT_i}{m_i}}$	0.1	n_e
0	$m_e + m_i$	0	0.01	$n_T(1 - r_i)$

Table 1: Argon plasma to analyse the collecting wall by a one-dimensional slabbing method.

being proportional to the separation between two consecutive slabs Δx. This understanding of the previous scheme ensures that the flux of f through the slab boundary must be positive, *i.e.*, $\mathbf{n} \cdot \mathbf{v} > 0$. Applying this approximation to equation (1), the resulting one-dimensional problem for each slab becomes

$$\frac{\partial f_\alpha}{\partial t} + \frac{q_\alpha E}{m_\alpha}\frac{\partial f_\alpha}{\partial v} = \left(\frac{\partial f_\alpha}{\partial v}\right)_{\alpha 0} + s_{flux}. \tag{5}$$

As initial condition we assume drift Maxwellian distributions, for each species of a quasi–neutral argon weakly ionized plasma, characterized by the parameters set on table 1. This initial condition ascertains that a large number of particles reaches the wall at $x = 0$, but still preserving an amount of returning ones to the quasi–neutral plasma, a fact due to collisional effects or to the action of the self–consistent electric field. The collision frequencies for the charge-neutral elastic exchange are $\nu^{\alpha 0} = \sigma^{\alpha 0} V_{th_\alpha}$, for constant cross–sections such as $\sigma^{i0} = 10^{-18}\,\mathrm{m}^2$ and $\sigma^{e0} = 10^{-20}\,\mathrm{m}^2$, where $V_{th_\alpha} = \sqrt{\frac{kT_\alpha}{m_\alpha}}$ is the species thermal velocity. A total density of particles $n_T = 10^{19}\,\mathrm{m}^{-3}$ and the ionization ratio $r_i = 10^{-6}$ are used. In this approach, charge-charge collisions are neglected due to their weak influence in the species dynamics for this low ionization ratio, as we argued in [28, Section 5.1.1]. Due to the upwind slabbing, a small Δx must be employed, which constrains the evolution to a reduced time step $\tau = 2 \cdot 10^{-5}$. This selection ensures that the upwind discretisation of the flux in the x direction is stable from a numerical point of view. This slightly perturbs the mesh-free basis of the PIM, but it is a requirement to ensure that the transversal flux is properly represented. A total of 200 slabs is considered to properly cover the space gap from $x = 0$ up to $x = 40\,\mathrm{cm}$.

3. Collecting wall

To simulate the collecting wall, a total absorbing boundary condition is imposed at the origin, therefore we set $f_\alpha(0, v > 0, t) = 0$. The distribution function for $f_\alpha(0, v \leq 0, t) = 0$ is described as a one-dimensional semi-open problem in the region $(-\infty, 0]$, by the boundary condition $f_\alpha(0, 0, t) = 0$. At $x = 40\,\mathrm{cm}$, the distribution function for negative velocities is kept constant to simulate a sustained plasma, meanwhile for positive velocities values this function is perturbed by the downstream flux and collisional processes. The density current, $J_\alpha(x, t) = [q_\alpha]$ is plotted against time, at positions $x = 0$ and $x = 40\,\mathrm{cm}$, in figure 1. Clearly, the wall acts as an initial perturbation that disrupts the quasi–neutral plasma. In figure 1b it can be seen how electrons move faster, creating an oscillatory situation that is dumped after a time of $1\,\mu\mathrm{s}$ approximately. After the initial perturbation, plasma adapts itself to the new situation, but each species evolve in its own rate, this is to say, the wall effect is quickly felt by the electrons, due to their high energy and low mass, in fact, it can be seen how the position far from the wall is influenced as positive velocity particles decrease, leading to a more positive electric current. This process leads the electrons to behave in an oscillatory mode to adapt themselves to the fast perturbation experienced by the wall. These oscillations, of unique frequency, are related to the approximated

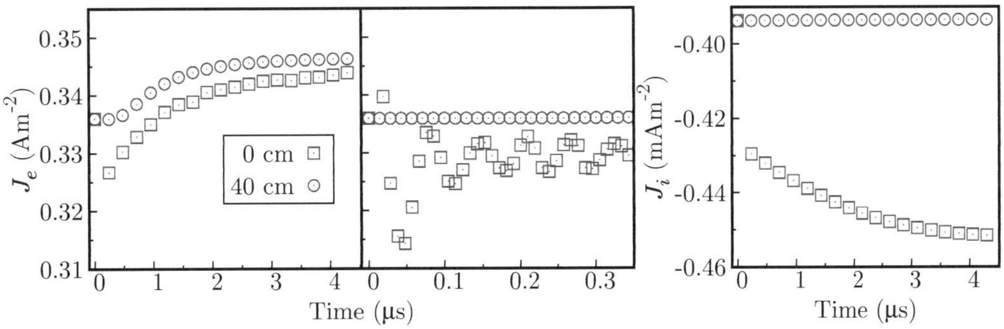

(a) Electrons (b) Initial perturbation (c) Ions

Figure 1: Density current for electrons and ions at $x = 0$ and $x = 40\,\mathrm{cm}$.

$f_{pe}(x = 0)$	$f_{pe}(x \to \infty)$	f_{pe} Numerical
18.4	20.1	17.0

Table 2: Electron plasma frequencies close to the collecting wall. All values are expressed in MHz.

cold plasma frequency f_{pe} of order

$$f_{pe} = \sqrt{\frac{n_e e^2}{\pi m_e}} \approx 8980\sqrt{n_e}\,\mathrm{Hz}, \tag{6}$$

where n_e is the electron density (all magnitudes are in cgs units). For this expression, ions have been assumed as infinitely massive. Table 2 presents the plasma frequency for three cases. Firstly, two cases corresponding to equation (6), using the electron density at $x = 0$ and far from the wall at time $t \simeq 0.15\,\mu\mathrm{s}$, are shown for the approximated values $4.2 \cdot 10^6$ and $5 \cdot 10^6\,\mathrm{cm}^{-3}$ respectively. The result obtained after evaluating the mean frequency for the cycles in figure 1b is presented. It can be seen how all frequencies are in the same order of magnitude of $100\,\mathrm{rad/s}$, but the numerical value is slightly lower than the analytical one for $x = 0$. Hence, it is reasonable to conjecture hat the electrons evolve in the simulations accomplishing a periodic dynamics close to the electron plasmas characteristic frequency. Moreover, this result arises naturally from the self–consistent resolution of the model, without imposing any constrain. The self–consistent resolution of the electric field is crucial to obtain this result, since small variations of this field are the main reason of the detected electron oscillatory behaviour. Therefore, we can assert that our simulations bring out an evolution that matches realistic physical plasma properties. Moreover, a perceptible evolution of ions is appreciated in figure 1c, meaning that the heavy species reacts to the perturbation produced by the wall, clearly in a larger time scale than the electrons one.

The figure 2 pictures the contours of f_e at four frame times. For short times, the effect of the wall only reaches close positions from it but, as time increases, the influence in the distribution function extends in space and an a significant amount of electrons is repelled by the metallic wall. As a result of this dynamics, we can conclude that the wall highly influences the distribution function for positive velocity, a fact that perturbs the quasi–neutral plasma. Although several works, as [34, 9], account for a non–homogeneous term to maintain the quasi–neutral plasma

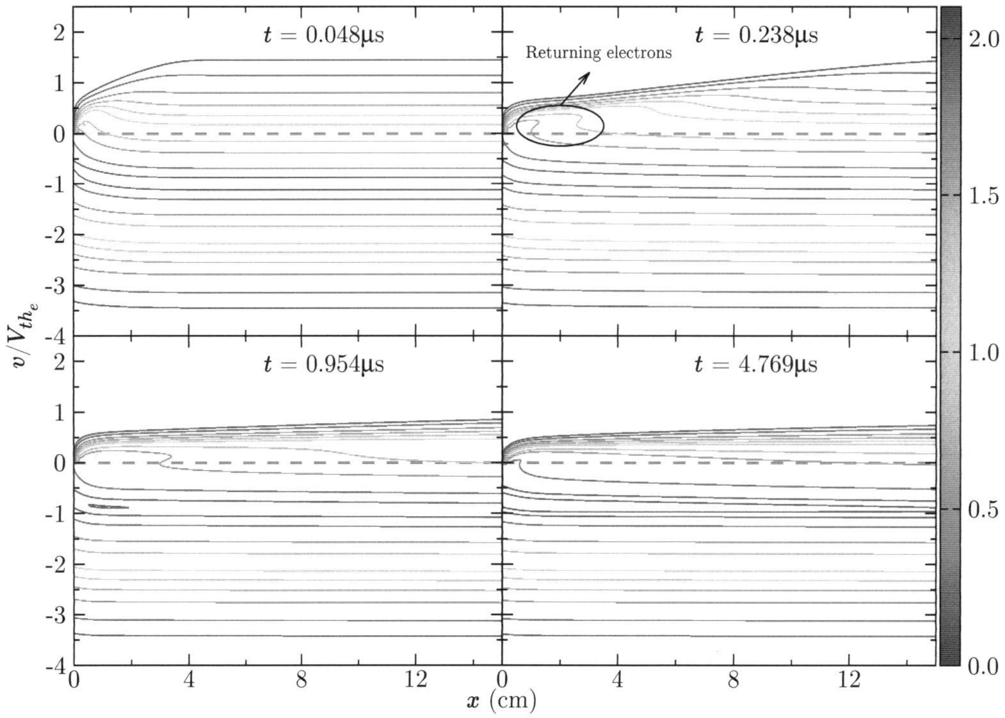

Figure 2: Contours of $f_e \times 10^{-7}$ in arbitrary units at different time frames. The relaxation of the returning electrons is clearly appreciated for $t < 1\,\mu s$

from a certain position, in the actual model this feature is achieved only by imposing the previously specified boundary conditions at $x = 40\,\text{cm}$. This means that all the domain, the region even far from the wall, can be affected by the presence of a metallic wall. In practice, it can be found that the use of an external source-sink of particles, to maintain the plasma, could perturb the system evolution [35] forcing a particular electric potential structure able o sustain the dynamics described here. Nevertheless, to properly establish the boundary conditions at 40 cm is still a difficult task that requires further research to describe the quasi-neutral distant plasma. For realistic models, enough space in the x direction should be provided to ensure that the system recovers a quasi–steady state without external terms, as it would correspond in a experimentally sustained plasma discharge. This would require a large number of plasma slabs and also a large computational time.

The boundary condition at $x = 0$ means that no particles are introduced into the numerical domain. However, it can be seen how a population of electrons still holds for $v > 0$. This is caused by the amount of electrons that return due to effect of the self–consistent electric field and the collision with neutrals. To analyse the first cause, the electric potential and field are pictured in figure 3. It can be seen how the fast evolution of electrons generates a large electric field due to the charge separation evolving to eliminate it by accelerating electrons positively and ions negatively. The electric field reduces its intensity as time increases and the potential difference between the wall and the end of the numerical domain is reduced drastically, as the plasma self–consistently adapts to the boundary condition. This creates some of the returning trajectories identified in figure 2.

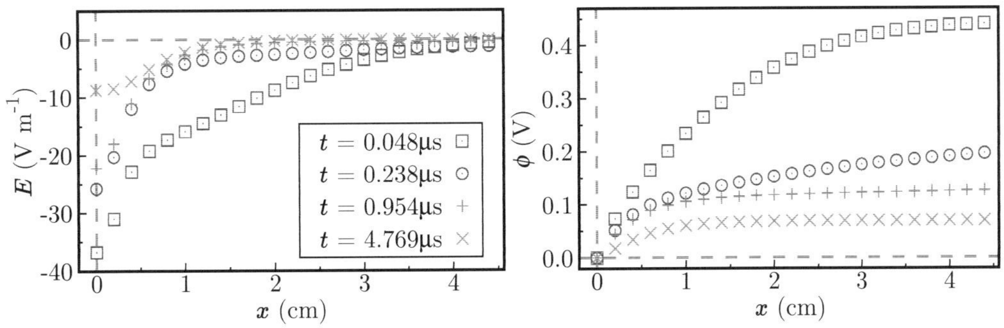

(a) Electric field

(b) Electric potential

Figure 3: **(a)** Electric field and **(b)** potential close to a collecting wall. Monotonic evolution is found for the potential, but interesting variations close to the wall arise for the electric field.

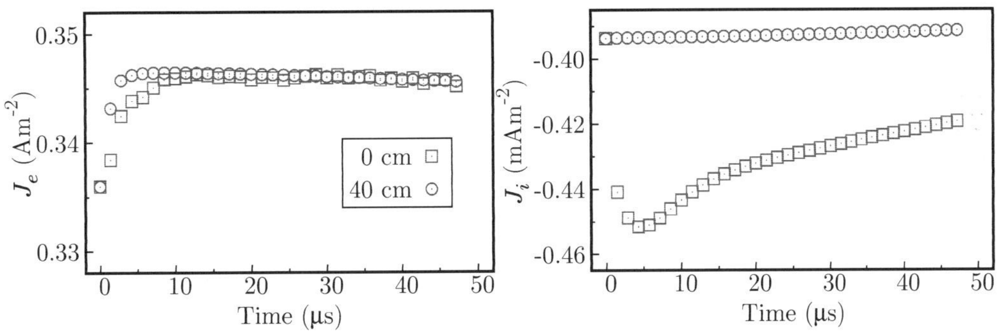

(a) Electrons

(b) Ions

Figure 4: Density current for **(a)** electrons and **(b)** ions for a time evolution of $\sim 50\,\mu$s. Electron density current at the wall and far from it coincide.

3.1. Long time analysis

In order to determine the influence of the ions on the system dynamics, it is interesting to analyse an extended period of time. The results presented above are analysed here for a period of time of order $50\,\mu$s in figure 4. The electron ensemble shows a fast evolution to a state in which the fluxes far and at the wall are equal at time $\sim 10\,\mu$s. In this time, the ions go through a fast change in order to fix the variations induced by the electron dynamics. Consequently, once the current at the wall and far from it become equal, the ions evolve under a slower time scale. However, the ion population seems to continue its evolution towards a situation in which, like the electrons, the current at the wall and far from it are equal.

Two differentiated regions in the system evolution can be distinguished, in a first one, up to a time of $\sim 15\,\mu$s, the system dynamics is guided by the fast electron response, which is the main cause of evolution of the self–consistent electric field. Then, electron flux is balanced and the small evolution of the ions is the dominant phenomena. Electrons also response to this variation as the current is slightly reduced for $t > 15\,\mu$s, as a response of the changes in the electric potential generated by the ions evolution.

4. Conclusions

The very involved and interesting problem of the plasma–wall interaction and related phenomena has been modelled and numerically solved by a slabbing kinetic model. This problem is of huge importance in the characterization of plasmas and its dynamics in bounded domains. Many disparate and odd processes determine the dynamics of these systems, as the distribution and energy of emitted electrons, the self–consistent electric field and the effects appearing in the vacuum chamber in which the plasma is held. Kinetic descriptions of these phenomena configure, in general, an arduous task, but every approach provides relevant information to understand the processes involved and how they influence the interaction. The aim of this procedures is to capture some behaviours that could be overleaped by some fluid plasma descriptions. For the models presented here, no assumption about the distribution function profile or the electric potential have been assumed, since they are obtained from a pure self–consistent kinetic problem we have stated.

In particular, the description of ions and electrons dynamics, in an argon weakly ionized plasma, in the presence of a collecting wall has been carried out. It has been found that the fluxes of the electrons at the wall and far from it became balanced, meaning that the plasma evolves to self-consistently equilibrate these flows. Moreover, despite the fact of the mass of ions is huge, compared with the electron mass, its dynamics close to a collecting wall modifies the electric field since ions are collected by the wall and its density is being reduced, giving rise to a smaller but important charge-separation. This is also appreciated for a large evolution time. Hence, a fast local change for the ions density occurs when the electrons tend to couple to the new situation, the ions evolve in a slow time scale and their dynamics can dominate the system evolution.

Physical results are obtained as a consequence of the self–consistent resolution of involved models, including the effect of the electric field or charge-neutral collisions. From this preliminary approach it is expected that future works can translate these models into cylindrical and spherical geometries to match the typical probes shapes to analyse the effects of probe geometries. More microscopic kinetic processes should be included to analyse their influence on close and far plasma dynamics. This also will require a more accurate description of the self–consistent electric field in the case of non-uniform wall emission model.

Acknowledgement

This work has been partially funded by the Ministerio de Economía Ciencia e Innovación of Spain under Grant ESP2013-41078-R.

References

[1] Sheehan J P, Hershkowitz N, Kaganovich I D, Wang H, Raitses Y, Barnat E V, Weatherford B R and Sydorenko D 2013 *Phys. Rev. Lett.* **111** 075002
[2] Tierno S P, Roibás E, Domenech-Garret J L, Donoso J M and Conde L 2013 *Contrib. Plasma Phys.* **53** 16–21
[3] Popov T K, Ivanova P, Dimitrova M, Kovačič J, Gyergyek T and Čerček M 2012 *Plasma Sources Sci. Technol.* **21** 025004
[4] Pierrard V and Lazar M 2010 *Solar Phys.* **267** 153–174
[5] Gyergyek T, Jurčič-Zlobec B, Čerček M and Kovačič J 2009 *Plasma Sources Sci. Technol.* **18** 035001
[6] Gyergyek T and Kovačič J 2015 *Physics of Plasmas* **22** 093511
[7] Gyergyek T and Kovačič J 2015 *Phys. Plasmas* **22** 043502
[8] Campanell M D and Umansky M V 2016 *Phys. Rev. Lett.* **116** 085003
[9] Campanell M D and Umansky M V 2017 *Phys. Plasmas* **24** 057101
[10] Sheehan J P, Kaganovich I D, Wang H, Sydorenko D, Raitses Y and Hershkowitz N 2014 *Phys. Plasmas* **21** 063502
[11] Chodura R 1982 *Phys. Fluids* **25** 1628–1633
[12] Devaux S and Manfredi G 2006 *Phys. Plasmas* **13** 083504
[13] Gonzalez J, Donoso J M and Conde L 2014 *Contrib. Plasma Phys.* **54** 298–303

[14] Chalise R and Khanal R 2015 *Phys. Plasmas* **22** 113505
[15] Tierno S P, Donoso J M, Domenech-Garret J L and Conde L 2016 *Phys. Plasmas* **23** 013503
[16] Ou J, Lin B and Zhao X 2017 *Phys. Plasmas* **24** 012510
[17] Gyergyek T and Kovačič J 2012 *Phys. Plasmas* **19** 013506
[18] Lieberman M A and Lichtenberg A J 1994 *Principles of plasma discharges and materials processing* (Cambridge Univ Press)
[19] Schrittwieser R, Balan P, Hron M, Ionita C, Jakubka K, Kryska L, Martines E, Stöckel J, Tichy M, Van Oost G *et al.* 2002 *Plasma Phys. Control. Fusion* **44** 567
[20] Sheehan J P and Hershkowitz N 2011 *Plasma Sources Sci. Technol.* **20** 063001
[21] Batani D, Alba S, Lombardi P and Galassi A 1997 *Rev. Sci. Instrum.* **68** 4043–4050
[22] Demidov V I, Ratynskaia S V and Rypdal K 2002 *Rev. Sci. Instrum.* **73** 3409–3439
[23] Dannenmayer K, Kudrna P, Tichý M and Mazouffre S 2011 *Plasma Sources Sci. Technol.* **20** 065012
[24] Raitses Y, Staack D, Smirnov A and Fisch N J 2005 *Phys. Plasmas* **12** 073507
[25] Roibás E, Tierno S P, Criado E, Domenech-Garret J L, Donoso J M and Conde L 2013 *Contrib. Plasma Phys.* **53** 57–62 ISSN 1521-3986
[26] Tierno S P, Domenech-Garret J L, Donoso J M, Jennewein D, Herdrich G, Fasoulas S and Conde L 2013 *IEEE Trans. Plasma Sci.* **41** 695–700 ISSN 0093-3813
[27] Criado E, Roibás E, Tierno S P, De Francisco P R, Domenech-Garret J L, Donoso J M and Conde L 2012 *Phys. Plasmas* **19** 023505
[28] Gonzalez J 2017 *Propagator Integral Method for Plasma Physics Kinetic equations with Practical Applications* Ph.D. thesis Escuela Técnica Superior de Ingeniería Aeronáutica y del Espacio
[29] Donoso J M and Gonzalez J 2017 Boundary-path-sum method for Drift-Diffusion equations in constrained domains under Review
[30] Matte J P and Virmont J 1982 *Phys. Rev. Lett.* **49**(26) 1936–1939 URL `https://link.aps.org/doi/10.1103/PhysRevLett.49.1936`
[31] Donoso J M and Salgado J J 2006 *J. Plasma Phys.* **39** 12587
[32] Gonzalez J, Donoso J M and Tierno S P 2016 *Phys. Plasmas* **23** 062311
[33] Dougherty J P and Watson S R 1967 *J. Plasma Phys.* **1** 317–326
[34] Gonzalez J 2013 *Design of a Semi-analytical Method to Describe Plasma-Wall Interactions* Master's thesis Escuela Técnica Superior de Ingenieros Aeronáuticos
[35] Campanell M D 2015 *Phys. Plasmas* **22** 040702

IOP Conf. Series: Journal of Physics: Conf. Series **958** (2018) 012005 doi:10.1088/1742-6596/958/1/012005

Determination of Anisotropic Ion Velocity Distribution Function in Intrinsic Gas Plasma. Theory.

A Mustafaev[1], A Grabovskiy[1], O Murillo[1] and V Soukhomlinov[2]
[1] Saint Petersburg Mining University, 21 line of the Vasilievskiy Island, h. 2 199106, Saint Petersburg, Russia. E-mail:schwer@list.ru
[2] Saint Petersburg State University, Universitetskaya, h. 7-9, 199034, Saint Petersburg, Russia. E-mail:prima-ivs@mail.ru

Abstract. The first seven coefficients of the expansion of the energy and angular distribution functions in Legendre polynomials for Hg^+ ions in Hg vapor plasma with the parameter $E/P \approx$ 400 V/(cm Torr) are measured for the first time using a planar one-sided probe. The analytic solution to the Boltzmann kinetic equation for ions in the plasma of their parent gas is obtained in the conditions when the resonant charge exchange is the predominant process, and ions acquire on their mean free path a velocity much higher than the characteristic velocity of thermal motion of atoms. The presence of an ambipolar field of an arbitrary strength is taken into account. It is shown that the ion velocity distribution function is determined by two parameters and differs substantially from the Maxwellian distribution. Comparison of the results of calculation of the drift velocity of He^+ ions in He, Ar^+ in Ar, and Hg^+ in Hg with the available experimental data shows their conformity. The results of the calculation of the ion distribution function correctly describe the experimental data obtained from its measurement. Analysis of the result shows that in spite of the presence of the strong field, the ion velocity distribution functions are isotropic for ion velocities lower than the average thermal velocity of atoms. With increasing ion velocity, the distribution becomes more and more extended in the direction of the electric field.

1. Introduction

The ion velocity distribution function (VDF) is interesting for many reasons among which there are problems associated with the analysis of plasma chemical reactions involving ions, determination of the ion mobility in a plasma object, heating of the neutral component of the plasma, etc. Among the technical applications one can mention modern plasma nanotechnologies, fine cleaning of surfaces of objects with ions and the technology of producing reliefs on surfaces by selective etching by means of bombardment with ion beams [1]. Peculiarities of ion drift were studied theoretically in [2–13], but the results are described most comprehensively in [5–13]. Perel [6] reported on the results of calculation of the drift velocity of ions of inert gases in their parent gas; however, the ion VDF was not calculated explicitly. The drift velocity of ions of inert gases in their parent gas plasma was also calculated in [5] in the strong field approximation; the author of [5] believed that the distribution of atoms has the form of a delta function. The author of [8], who determined the distribution function of ions in their parent gas, disregarded the generation of ions with a Maxwellian velocity distribution as a result of charge exchange. It will be shown below that this may lead to considerable errors in the ion VDF.

The time during which an ion moving in its parent gas (if we disregard all processes except charge exchange) has a velocity component in the direction of the electric field in the interval from v_{iz} to $v_{iz} + dv_{iz}$ [7]. Clearly, this time is proportional to the ion distribution function of the velocity projection onto the electric field direction; however, it is difficult to obtain from these results the total velocity distribution function.

In [9, 10], a new method for calculating matrix elements of the collision integral was developed, which was used for solving the time-dependent Boltzmann equation by the method of moments for ions in the conditions when the resonant charge exchange is the predominant process. However, the application of this method for calculating the time-independent VDF for ions is hampered in strong field, when the ratio of the thermal energy of atoms and the energy of an ion acquired over the mean free path is less than 0.05.

In [11], the analytic solution to the Boltzmann equation for ions in the Bhatnagar–Gross–Krook (BGK) model [14] is compared with the numerical solution for a constant resonant charge-exchange cross section. It is shown that in strong fields, the analytic solution gives an incorrect asymptotic form for the drift velocity of an ion.

The author of [12] developed the moments method for solving the Boltzmann equation for charged particles for different potentials of their interaction with atoms.

In [13], a review of the methods for solving the Boltzmann equation for charged particles is given both for the regimes close to hydrodynamic conditions and in the opposite case. In the latter, the principles of the moment method (and its modifications) are given. The cases of both constant and varying fields (in particular, RF discharge) are considered.

As regards the measurement of the VDF for ions in the plasma of a self-consistent dc discharge, we are not aware of publications except [15], in which the Doppler shift of the ion lines is measured spectroscopically for a discharge in argon and observating along the discharge axis. From this shift, the average velocity of ions was estimated, which was found to be 10^4 cm/s in the given experimental conditions.

Various aspects of ion drift in the plasma of lowpressure capacitive discharge are studied theoretically in [16–18].

This study is devoted to experimental and theoretical determination of the ion VDF in their parent gas of the glow discharge taking into account the generation of slow ions with the atomic temperature as a result of charge exchange, which was treated as the dominating process. We assumed that the velocity of ions prior to collisions considerably exceeds the velocity of atoms.

2. Theoretical determination of the ion distribution function

2.1. Derivation of Basic Relations for a Constant Cross Section in Zero Ambipolar Field

The mechanisms of the interaction of ions with remaining components of a low-temperature plasma have been studied comprehensively [2, 19, 20]. It was found that the dominating process in collisions of an ion in its parent gas is resonant charge exchange as a result of which an ion colliding with an atom transfers to it its charge due to the tunnel effect, but preserves the energy and momentum. Thus, the motion of the ion in its parent gas is a relay-race process (Sena effect [2]).

Considering the main mechanism of interaction of ions with neutral atoms, we analyze the steady-state velocity distribution under the following conditions: (i) gas atoms move in accordance with the Maxwell law; (ii) ions move in their parent gas with a low degree of ionization; (iii) the dominating process that forms the ion velocity distribution function is resonant charge exchange; (iv) longitudinal gradients of the plasma parameters are equal to zero. The last condition is necessary because otherwise the ion VDF will also depend on the longitudinal coordinate. The solution to the problem in this case can also be obtained easily, but will require the imposition of physically substantiated boundary conditions along the *z* coordinate.

In a stationary plasma under the above assumptions, the Boltzmann equation has the form:

$$\vec{v}_i \vec{\nabla}_r (f_i) + \frac{eE}{m} \vec{\nabla}_v (f_i) = S_i, \qquad (1)$$

where e and m are the ion charge and mass, $\vec{E}$ is the electric field strength, $\vec{v}_i$ is the velocity of ions, f_i is the ion VDF, and S_i is the collision integral. Considering that an ion appearing as a result of charge exchange as well as a result of an electron impact has the velocity of the atom, we define the collision integral as

$$S_i(\vec{v}_i) = n_a \iint \sigma v_r f_a\left(\vec{v'}_a\right) f_i\left(\vec{v'}_i\right)\left[\delta\left(\vec{v'}_a - \vec{v}_i\right) - \delta\left(\vec{v'}_i - \vec{v}_i\right)\right] d\vec{v'}_a d\vec{v'}_i, \tag{2}$$

where $f_a(\vec{v}_a)$ is the atomic distribution function, n_a is the concentration of atoms, σ is the charge-exchange cross section, v_r is the magnitude of the relative velocity of ion–atom, and $\vec{v}_a$ is the velocity of atoms. In the collision integral, we disregard the ionization and electron–ion recombination processes since under the given assumptions, their characteristic time is large as compared to the resonant charge-exchange time. The ion distribution function is normalized to the concentration and the distribution function for atoms is normalized to unity.

First of all, let us specify the functions appearing in the collision integral. It is well known that the charge exchange cross section is a logarithmic function of the ion–atom relative velocity [5, 20, 21]:

$$\sigma(v) = \sigma_0\left[1 + a \cdot ln\left(\frac{v_0^2}{v^2}\right)\right] \equiv \sigma_0 \kappa(v), \tag{2a}$$

where σ_0, a, v_0 are the constants depending on the type of the gas. For example, $\sigma_0 = 2.79 \cdot 10^{-15}\ cm^2$ and a = 0.0557 for helium with $\frac{mv_0^2}{2} = 1\ eV$ [21], while $\sigma_0 = 12 \cdot 10^{-15}\ cm^2$ and a = 0.073 for resonant charge exchange of Hg^+ ions [22]. At the same time, the author of [23] mentioned the following values: $\sigma_0 = 20 \cdot 10^{-15}\ cm^2$ and a = 0.048. In view of such a weak energy dependence in a comparatively narrow velocity range typical of the motion of ions, the charge-exchange cross section can be treated as constant with a high degree of accuracy. Indeed, in the energy range 0.1– 0.3 eV, the charge-exchange cross section (e.g., for helium) changes by approximately 5% according to the results from [9, 21].

For this reason, we will first solve the problem assuming that the resonant charge-exchange cross section is constant and then write the formulas taking into account its dependence on velocity. Let us consider the conditions of motion of ions in which the velocity acquired by them over the mean free path is considerably higher than the average velocity of atoms; such conditions hold if $\sqrt{kT_a} \ll \sqrt{\frac{eE}{\sigma n_a}}$ (T_a is the temperature of neutral atoms). This inequality is equivalent to

$$(E \cdot P^{-1} \cdot \sigma^{-1})^{1/2} \gg 1, \tag{2b}$$

where P [Torr] is the gas pressure, E is measured in [V/cm], and σ [$10^{-15}\ cm^2$].

It should be noted that in the case of inert gases, inequality (2b) holds only for small values of the product $PR \ll 1$ (R is the characteristic size of the volume in which the plasma is produced), when the electric field in the plasma increases substantially due to the diffusion induced by the perishing of electrons [24]. In this case, we assume that the ion–atom relative velocity prior to the collision has only one component along the electric field strength $\vec{E}$ and is determined only by the velocity of the ion. We also assume that the atomic collision frequency is considerably higher than the frequency of ion–atom collisions. This allows us to use the Maxwell equilibrium distribution for atoms, but imposes a limitation on the degree of ionization. Under the above assumptions, the collision integral can be reduced to the expression:

$$S_i(\vec{v}_i) = n_a\sigma[f_a(\vec{v}_i)\int|v_{iz}|f_i(\vec{v}_i)d\vec{v}_i - |v_{iz}|f_i(\vec{v}_i)]. \quad (3)$$

Let us first consider the case when the plasma is not confined by the walls and the electric field has one component directed along the z axis. We assume that this field component, as well as the concentrations of atoms and ions, is independent of the coordinate. Then, distribution function f_i depends only on velocity $\vec{v}_i$ and we can, using expression (3), reduce the kinetic equation to the following form up to terms of the order $\left(\frac{\alpha_0}{\beta}\right)$:

$$\frac{\partial f_i}{\partial v_{iz}} + 2\alpha_0|v_{iz}|f_i = \gamma f_a(\vec{v}_i), \quad (4)$$

where $O(x)$ is a quantity of the same order of magnitude as x for $x \to 0$; $2\alpha_0 = \frac{m}{eE}n_a\sigma$; $\gamma = 2\alpha_0 n_i \bar{v}_{iz}$; $f_a(\vec{v}_i) = \left(\frac{\beta}{\pi}\right)^{1.5} exp(-\beta v_i^2)$; $\beta = \frac{m}{2kT_a}$; T_a - is the temperature of atoms, $\bar{v}_{iz} = \frac{\int v_{iz} f_i(\vec{v}_i)d\vec{v}_i}{\int f_i(\vec{v}_i)d\vec{v}_i}$ is the average velocity of ions in the direction of the electric field, and n_i is the ion concentration.

Let us now assume that the plasma is produced in a cylindrical discharge tube of radius R with dielectric walls and the possibility of the plasma being heated along the radius non-uniformly. In this case, the left hand side of Eq. (4) acquires three additional terms $\frac{m}{eE}v_\rho\frac{\partial f_i}{\partial \rho}$; $\frac{E_\rho(\rho)}{E}\frac{\partial f_i}{\partial v_\rho}$; $\frac{m}{eE}\frac{v_\rho}{n_i}\frac{\partial n_i}{\partial \rho}f_i$, where E_ρ is the ambipolar field, ρ is the radial coordinate, and v_ρ is the ion velocity component in the radial direction. The last term is of the order of $\frac{m}{eE}\frac{v_\rho}{R}f_i \ll \frac{m}{eE}\frac{v_{iz}}{\lambda_i}f_i$ for $\lambda i \ll R$, where λ_i is the path length of an ion in respect with the resonant charge exchange, and therefore can be disregarded. The second term is much smaller than $\frac{\partial f_i}{\partial v_{iz}}$ if the inequality $E_\rho(\rho) \ll E$ holds in the self-sustained plasma discharge. This inequality is observed near the discharge axis (where the total ambipolar field is zero) or in the entire volume if the following relation holds [24]:

$$PR \gg 10^{-3} \cdot T_a \cdot \sigma_e^{-1} \cdot \delta^{-1/2}, \quad (4a)$$

where σ_e is the elastic scattering cross section of an electron from a neutral particle and δ is the average relative fraction of the electron energy lost by in the elastic scattering from neutral particles of the plasma; $P\ [Torr]$, $R\ [cm], T_a\ [K], \sigma_e\ [10^{-16}\ cm^2]$.

The first additional term is obviously of the order of the last term; therefore, all three terms can be disregarded. If inequalities (4a) and $\lambda_i \ll R$ hold, in the case of a nonuniform heating of the gas, the $f_i(\rho)$ dependence can be taken into account just assuming that quantities α_0, β, γ are functions of the radius.

Let us first assume that these inequalities hold and parameters α_0, β, γ that determine the properties of the plasma are specified a priori (including their possible dependence on radial coordinate ρ). It should be noted that parameter α_0 determines the velocity acquired by an ion in the electric field, parameter β characterizes the temperature of neutral atoms and γ is the rate of formation of ions due to charge exchange.

The solution to kinetic equation (4) when inequalities (4a) and $\lambda_i \ll R$ take place (inequality (4a) for small $\frac{\rho}{R}$ holds because $E_\rho(0) = 0$)), taking into account the behavior of the ion velocity distribution function for $v_{iz} \to -\infty$, has the form

$$f_i(\vec{v}_i) = Aexp[-\beta v_i^2 + (\beta - \alpha_0)v_{iz}^2]erfc(-\sqrt{\beta - \alpha_0}v_{iz})\left\{1 + O\left(\frac{\alpha_0}{\beta}\right)\right\} \text{ при } v_{iz} \geq 0;$$

$$f_i(\vec{v}_i) = Aexp[-\beta v_i^2 + (\beta + \alpha_0)v_{iz}^2]erfc(-\sqrt{\beta + \alpha_0}v_{iz})\left\{1 + O\left(\frac{\alpha_0}{\beta}\right)\right\} \text{ при } v_{iz} \leq 0, \quad (5)$$

where $A = \frac{\gamma\beta}{2\pi}$ is a constant factor, v_i is the magnitude of the ion velocity, and $erfc(x) = \frac{2}{\sqrt{\pi}}\int_x^\infty \exp(-t^2)\,dx$. Under the above assumptions, the inequality $\sqrt{\frac{\alpha_0}{\beta}} \ll 1$ holds. Note that expression (5) for $v_{iz} = v_i\mu$; $\mu = cos\theta$ (θ is the angle between the *z* axis and the ion velocity) gives the angular velocity distribution for ions. Integrating the relation (5) with respect to angular coordinates, we obtain the ion distribution function over velocity modulus, normalized to unity:

$$F_i(v_i) \equiv \frac{f_i(v_i)}{n_i}v_i^2 = \frac{2\sqrt{\alpha_0\beta}}{\sqrt{\pi}}v_i exp(-\beta v_i^2)\int_{-\sqrt{\beta+\alpha_0}v_i}^{\sqrt{\beta-\alpha_0}v_i} exp(t^2)erfc(-t)dt\cdot\left\{1+O\left(\frac{\alpha_0}{\beta}\right)\right\}. \quad (6)$$

The ion concentration can be expressed in terms of the constants defined above as follows: $n_i = \frac{\gamma\sqrt{\pi}}{2\sqrt{\alpha_0}}$. It should be noted that to take into account terms $\left(\frac{\alpha_0}{\beta}\right)$, it is sufficient to specify the normalization in this formula i.e., divide the function by $\int_0^\infty F_i(v_i)\,{v_i}^2 dv_i$.

Integration of relation (5) with respect to the azimuthal angle and the velocity component orthogonal to the electric field direction gives the distribution of ions over the *z* component of velocity v_{iz}, normalized to the concentration:

$$f_{iz}(\vec{v}_i) = \frac{\pi A}{\beta}exp(-\alpha_0 v_{iz}^2)erfc(-\sqrt{\beta-\alpha_0}v_{iz})\left\{1+O\left(\frac{\alpha_0}{\beta}\right)\right\} \text{ при } v_{iz} \geq 0; \quad (6a)$$

$$f_{iz}(\vec{v}_i) = \frac{\pi A}{\beta}exp(\alpha_0 v_{iz}^2)erfc(-\sqrt{\beta+\alpha_0}v_{iz})\left\{1+O\left(\frac{\alpha_0}{\beta}\right)\right\} \text{ при } v_{iz} \leq 0.$$

The ion energy distribution function (IDF) $F_{i\varepsilon}(\varepsilon_i)$, normalized to unity can be written (with the same accuracy as relation (6)) in the form:

$$F_{i\varepsilon}(\varepsilon_i) = \frac{1.16\cdot10^4\sqrt{\alpha_0}}{\sqrt{\pi}T_a}exp[-E(\varepsilon_i)]\int_{-\sqrt{\left(1+\frac{\alpha_0}{\beta}\right)E(\varepsilon_i)}}^{\sqrt{\left(1-\frac{\alpha_0}{\beta}\right)E(\varepsilon_i)}} exp(t^2)erfc(-t)dt\left\{1+O\left(\frac{\alpha_0}{\beta}\right)\right\}, \quad (6b)$$

where ε_i is the ion energy in electron volts, $E(\varepsilon_i) = \frac{1.16\cdot10^4\varepsilon_i}{T_a}$, terms $O\left(\frac{\alpha_0}{\beta}\right)$ are taken into account analogously to formula (6) (i.e., by specifying the renormalization), and the azimuthal angle and energy IDF have the form:

$$F_{i\varepsilon\mu} = \frac{1.16\cdot10^4\sqrt{\alpha_0}}{\sqrt{\pi}T_a}\sqrt{E(\varepsilon_i)}exp\left[-E(\varepsilon_i)(1-\mu^2)-\frac{\alpha_0}{\beta}E(\varepsilon_i)\mu^2\right]erfc\left[-\sqrt{1-\frac{\alpha_0}{\beta}E(\varepsilon_i)}\mu\right] \quad (6c)$$

for $\mu > 0$

$$F_{i\varepsilon\mu} = \frac{1.16\cdot10^4\sqrt{\alpha_0}}{\sqrt{\pi}T_a}\sqrt{E(\varepsilon_i)}exp\left[-E(\varepsilon_i)(1-\mu^2)+\frac{\alpha_0}{\beta}E(\varepsilon_i)\mu^2\right]erfc\left[-\sqrt{1+\frac{\alpha_0}{\beta}E(\varepsilon_i)}\mu\right]$$

for $\mu < 0$.

It follows from the expression for n_i and from definition of parameters α_0 and γ that the average velocity of an ion is defined to within terms of the order of $\frac{\alpha_0}{\beta}$ as:

$$\bar{v}_i = \sqrt{\frac{1}{\pi\alpha_0}}\left[1+O\left(\frac{\alpha_0}{\beta}\right)\right] \approx \sqrt{\frac{2eE}{\pi n_a\sigma}}, \quad (7)$$

The same result can be obtained by evaluating the average velocity directly using the distribution function obtained above. The calculation of the square mean velocity with the same degree of accuracy gives

$$\sqrt{\bar{v}_i^2} = \left(\frac{1}{n_i}\int_0^\infty F_i(v_i)\,dv_i\right)^{0.5} = \frac{1}{\sqrt{2\alpha_0}}\left[1 + O\left(\frac{\alpha_0}{\beta}\right)\right], \tag{7a}$$

while for the most probable velocity v_{ip} in the approximation $\alpha_0 \to 0$ we find from the solution of the transcendental equation $\left(\frac{dF_i(v_i)}{dv_i}\right)_{v_i=v_{ip}} = 0$, that $v_{ip} = \frac{1.5}{\sqrt{\beta}}$. The allowance for the finite value of parameter $\alpha_0 \ll \beta$ shifts the position of the peak of the ion VDF towards lower velocities.

2.2. Inclusion of the Ambipolar Field and Dependence of the Charge-Exchange Cross Section on Velocity

In the case when the ambipolar field must be taken into account (i.e., if we seek the solution not in the axial region of the discharge or if inequality (4a) does not hold) but the inequality $\lambda_i \ll R$ still takes place, we proceed as follows. Taking into account the fact that we can disregard in our case the spatial gradients in the Boltzmann equation and that the resonant charge exchange is the main process determining the form of the distribution function, we will solve the problem for a certain real coordinate ρ by choosing a local system of coordinates in which the *z* axis coincides with the direction of the total electric field in the plasma for this value of ρ:

$$E_s = E\sqrt{1+\varepsilon(\rho)^2}, \tag{8}$$

where $\varepsilon(\rho) = \frac{E_\rho(\rho)}{E}$.

Then, the Boltzmann equation has the form (4), but the parameter α_0 depends on the ratio of the ambipolar and axial (directed along the discharge tube axis) fields:

$$\alpha(\rho) = \alpha_0\sqrt{1+\varepsilon(\rho)^2}. \tag{8a}$$

Respectively, expressions (5) and (6) hold after the substitution of α_0 for the parameter $\alpha(\rho)$. In addition, a simple geometrical analysis shows that we must replace $v_z = v\mu$ in formula (5) by the expression $\frac{v\left[\mu+\varepsilon(\rho)\sqrt{1-\mu^2}\right]}{\sqrt{1+\varepsilon(\rho)^2}}$, and $v_\rho = v\sqrt{1-\mu^2}$ by $\left|\frac{v\left[\varepsilon(\rho)\mu-\sqrt{1-\mu^2}\right]}{\sqrt{1+\varepsilon(\rho)^2}}\right|$. It should be noted that we have not imposed any constraints on the relation between the axial and ambipolar fields. As regards the expression for the drift velocity in the axial direction, taking into account the fact that the relation

$$\bar{v}_i = \bar{v}_{iz}\sqrt{1+\varepsilon(\rho)^2}, \tag{9}$$

holds in the chosen system of coordinate, we can write

$$\bar{v}_{iz} = \sqrt{\frac{2eE}{\pi n_a \sigma}} \cdot \frac{1}{[1+\varepsilon(\rho)]^{0.25}}. \tag{10}$$

Concluding this section, let us consider how the resultant formulas will change if we take into account the dependence of the resonant charge-exchange cross section on the velocity. Calculations show that the values of quantities $\frac{x^2\kappa(x)}{2}$ *and* $\int_0^x \kappa(y)ydy$, (e.g., for inert gases) change with the variation of the quantity $x = \sqrt{\beta}v_i$ in the range $x \in [0, 10]$ by not more than 5%, where $\kappa(x)$ is defined by formula (2a). We can easily show that all expressions derived above describe to the same degree of accuracy the corresponding quantities in the case of dependence of the charge-exchange cross section on the velocity in accordance with expression (2a) after the substitution of $\alpha(\rho)$ for $\alpha(\rho)\kappa(v_i)$.

The problem was solved under the assumption that inequality $\lambda_i \ll R$ holds. This is necessary for disregarding the term with $\vec{\nabla}_r(f_i)$ in kinetic equation (1), However, even if this inequality does not hold, the above results are still valid in the vicinity of the tube axis because $\vec{\nabla}_r(f_i)_{r=0} = 0$. We can easily prove that for an arbitrary ratio , the exact solution to the Boltzmann equation tends to the solution to Eq. (4) for $\rho \rightarrow 0$.

3. Experimental determination of the ion distribution function in plasma glow discharge

We measured the energy IDF by the probe method in the positive column of the low-pressure glow discharge in mercury. The discharge was initiated in a quartz glass tube 30 mm in diameter and 300 mm in length between the planar impregnated indirectly heated cathode 11 mm in diameter and a molybdenum anode with a diameter of 20 mm. Preliminary thermal preparation of the device and oil-free evacuation ensured a vacuum of 10^{-8} Torr in the working regime.

The cathode temperature was measured by a W– Re microthermocouple and was maintained by the electron stabilization system at a level of 1000–1500 K to within ±5 K.. The stability of the emission cathode current and the discharge current during measurements at various mercury pressures were controlled by the electronic stabilization system.

Mercury was fed to the tube using the vacuum distillation method. The saturated vapor pressure was controlled by the temperature of the thermostat containing the liquid phase of mercury. The mercury pressure varied in the range $5\cdot10^{-4}-1$Torr and was monitored by magnetodischarge sensors. The upper boundary of the pressure was determined by the sizes of probes and the applicability conditions for the collisionless theory of the probe current. The discharge current varied from 0.05 to 0.5 A. The estimates of the near-probe layer thickness show that the collisionless layer approximation [25, 26] holds in experimental conditions, and the thickness of this layer is much smaller than the probe diameter.

The main experimental difficulty in the IDF measurements with a planar probe is that it must rotate and simultaneously move along the discharge axis. To overcome this difficulty, a rotating planar disk-shaped one-sided probe made of 30-μm-thick tantalum foil 0.5 or 0.8 mm in diameter is introduced into the plasma through the lateral boundary. The probe is located on the discharge tube axis. A tantalum wire lead of diameter 0.1 mm is welded to the probe. The lead and one side of the probe are protected with a special alundum coating annealed in vacuum. The probe was mounted in accordance with a three-coordinate micrometric remote displacement system, which ensured its spatial fixation to within ±0,01mm and orientation relative to the symmetry axis of the discharge in the angular range 0–180° with a step of 5° with accuracy not worse than ±10″. The installation of the probe was monitored with the help of an ocular micrometer.

Legendre coefficients F_{ia}^n of the IDF were measured by recording the second derivatives of the probe current I''_U, obtained by double modulation of the probe potential. The experimental setup and the method for measuring the second derivative of the probe current with respect to the probe potential were described in detail in [27, 28]. We used a modulating signal of the form $u(t) = \Delta\varepsilon(1 + cos\omega_1 t)cos\omega_2 t$.

The consideration of hardware distortions, the choice of the optimal amplitudes of the differentiating signal $\sqrt{2}u_1$ and the control over the recording system linearity were carried out in accordance with the technique described in [29, 30]. We used the following values: $\Delta\varepsilon$ = 0.05, 0.1, and 0.2 V, $\omega_1 = 6 \cdot 10^3\ Hz$ *and* $\omega_2 = 6 \cdot 10^5\ Hz$.

In the range of negative probe potentials, the value of I_U'' is determined by the electron component of the probe current and is proportional to the electron VDF. It is well known [31] that for a positive probe potential, the electron current to the planar probe is almost independent of this potential. Ions move in this case in the retarding field, and the second derivative of the ion current carries information on their distribution function. For this reason, I_U'' in the potential range 0–1 V is mainly describes the ion distribution. Special attention in recording I_U'' was paid to the stability of the discharge current.

In our experiment, we tried to eliminate distortions emerging during IDF measurement due to the finite conductivity of plasma [32, 33]. For this purpose, a planar one-sided probe and a spherical probe of diameter 0.3 mm were soldered into the tube at distances of 20 and 25 cm from the cathode, respectively, for monitoring the accuracy in the reconstruction of the isotropic component F_{ia}^{0} of the IDF. The absence of distorting effect of vibrations in measurements was monitored by recording I_U'' in the absence of a differentiating signal.

The special measures taken in our experiment for stabilizing the electric discharge parameters ensured the reproducibility of the results not worse than 0.5%. The oscillation spectrum of the discharge current and the voltage were monitored in a frequency range up to 300 MHz.

The special calibration of the measuring system ensured not only relative, but also absolute measurements of I''_U, which in turn made it possible to determine angular harmonics of the distribution function F_{ia}^{n} in an absolute way.

The spatial potential was determined as the zero of the second derivative of the probe current [26]. The electron and ion concentrations were calculated by integrating the distribution functions for electrons and ions. Comparison of these concentrations shows their coincidence to within 10%, which is a satisfactory result considering the error of probe measurements.

The experiment was controlled with help of a PC based multichannel measuring–computational system. Special software and radio-engineering elemental base made it possible to perform digital recording of the data obtained in pulse and stationary regimes and complex processing in real time scale.

Digital recording made it possible to substantially improve the accuracy and sensitivity of the diagnostic method and elevated the authenticity of experimental results.

4. Discussion of results and comparison of experimental and computation data

It should be noted that expression (7) for the ion drift velocity coincides with the results obtained in [6, 19] in the limit case of a strong field. Analogously, formula (10) coincides with the result reported in [15]. As mentioned above the time during which an ion has a velocity component along the electric field in the interval from v_{iz} to $v_{iz} + dv_{iz}$was calculated in [7]. This time determines the IDF for the *z* component of velocity. Distribution (6a) obtained in this study coincides with expressions (47) (for $v_{iz} \geq 0$) and (46) (for $v_{iz} \leq 0$) from [7] for $\frac{\beta}{\alpha_0} \gg 1$. When the relation $\sqrt{\frac{\beta}{\alpha_0}} \to \infty$ takes place, distribution (6a) is transformed to the Maxwellian distribution identical to those obtained in [8]. As noted above, this corresponds to the disregard of ions generated as a result of charge exchange with thermal velocities. Thus, we can state that the result obtained here for the IDF are in conformity with the results of other authors.

Let us now consider some peculiarities of distribution function (6). It can be seen that the ion velocity modulus distribution function depends on two parameters, α_0 and β, which distinguishes this distribution from the well-known equilibrium Maxwellian distribution.

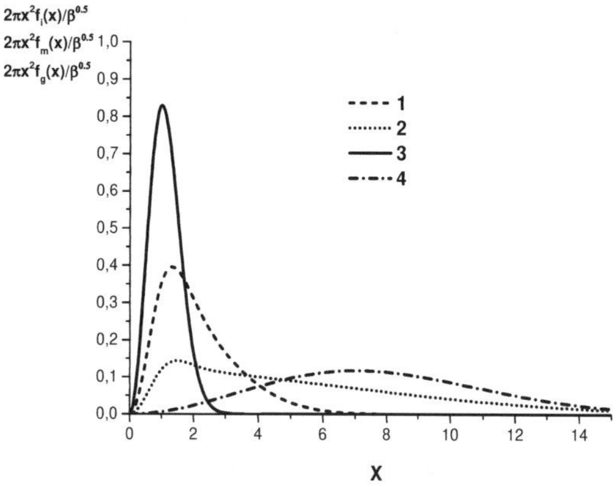

Figure 1. Distribution function $x^2 f_m(x)$ for ions and $x^2 f_i(x)$, and $x^2 f_g(x)$ for ions (see formula (11) from [8]); $x = \sqrt{\beta} v_i$, (*1*) $x^2 f_i(x)$, for $\frac{\alpha_0}{\beta} = 0.1$; (*2*) $x^2 f_i(x)$, for $\frac{\alpha_0}{\beta} = 0.01$, (*3*) $x^2 f_m(x)$; and (*4*) $x^2 f_g(x)$ for $\frac{\alpha_0}{\beta} = 0.01$.

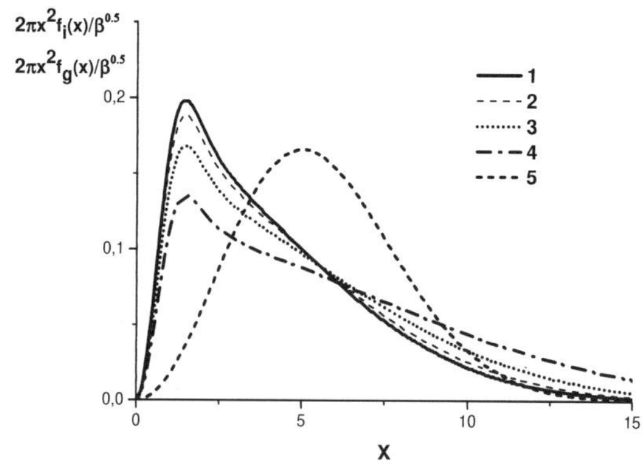

Figure 2. Distribution functions $x^2 f_i(x)$ and $x^2 f_g(x)$ for $\frac{\alpha_0}{\beta} = 0.02$ for different ratios ε(ρ) of the ambipolar field to the axial field: (*1*) $x^2 f_i(x)$ for ε(ρ) = 0; (*2*) $x^2 f_i(x)$ for ε(ρ) = 0.5; (*3*) $x^2 f_i(x)$ for ε(ρ) = 1; (*4*) $x^2 f_i(x)$ for ε(ρ) = 2; and (*5*) $x^2 f_g(x)$.

Figure 1 shows the curves describing functions $x^2 f_i(x)$, where $x = \sqrt{\beta} v_i$, for various values of the parameter $\frac{\alpha_0}{\beta}$ according to (6), of the Maxwellian distribution $x^2 f_m(x) = \frac{4}{\sqrt{\pi}} x^2 exp(-x^2)$and the distribution obtained in [8] for the constant resonant charge-exchange cross section:

$$x^2 f_g(x) = \frac{8\sqrt{2}}{\sqrt{\pi}} \left(\frac{\alpha_0}{\beta}\right)^{1.5} x^2 exp\left(-\frac{2\alpha_0 x^2}{\beta}\right). \tag{11}$$

It can be seen that the number of particles for function $x^2 f_i(x)$ in the range of high velocities is much larger than for the equilibrium distribution with atomic temperature. The opposite relation in observed in the range of low velocities. This can easily be explained by the fact that the excess of ions in the range of high velocities is associated with their acceleration by the electric field, while the deficit of ions in the range of low velocities as compared to the Maxwellian distribution is due to persistent departure of ions from this region as a result of the same acceleration in the electric field. At the same time, the departure of particles in the range of low energies in the case of the equilibrium distribution is a discrete process that occurs only during their collisions with other particles (some of the collisions lead to energy loss).

It should also be noted that the group of slow ions forming the ion VDF in the region of its peak has the characteristic velocity on the order of the atomic velocity. However, this does not mean that the assumption that the ion velocity prior to the collision must be much higher than the thermal velocity of atoms is violated for this group of ions. Indeed, the group of ions with thermal velocities appears precisely as a result of charge exchange; i.e., these are the ions that have just been generated and that must traverse before the next collision a distance of the order of the path length over which they are accelerated in the field to high (as compared to thermal) velocities.

Comparison of distributions (6) and (11) (Figs. 1, 2) shows that these distributions differ radically. This is due to the difference in the physical models used for them. As mentioned above, the author of

[8] disregarded the generation of ions as a result of charge exchange assuming that this can be done since the velocity of an ion generated in this way is much lower than the average ion velocity (acquired due to the field over the path length) in accordance with the conditions of the problem. This is true indeed, but the concentration produced by the ion accelerated in the field in the case of equal fluxes is disregarded. To clarify this situation, it is sufficient to consider the 1D problem of motion of a charged particle beam in the field. In the stationary case, since the flux is constant, the concentration decreases upon an increase in the velocity. It is for this reason that the most probable velocity (obtained in this study) is determined by the temperature of atoms (i.e., ions that have just been generated).

It can be seen from results (7) and (7a) that upon an increase in the ambipolar field, the average and square mean velocities of ions become higher, which is quite clear because ions are additionally accelerated in the ambipolar field. At the same time, the average velocity along the axial field decreases in accordance with the relation (10). This is completely confirmed by the data depicted in Fig. 2, which shows IDFs for various parameters ε(ρ). The same figure shows distribution function $x^2 f_g(x)$ [8]. It is shown above that the most probable velocity of ions has the same order of magnitude as the most probable velocity of neutral atoms and is determined by parameter β. On the other hand, average and square mean velocities are determined by parameter $\alpha(\rho)$ in accordance with expressions (7) and (7a).

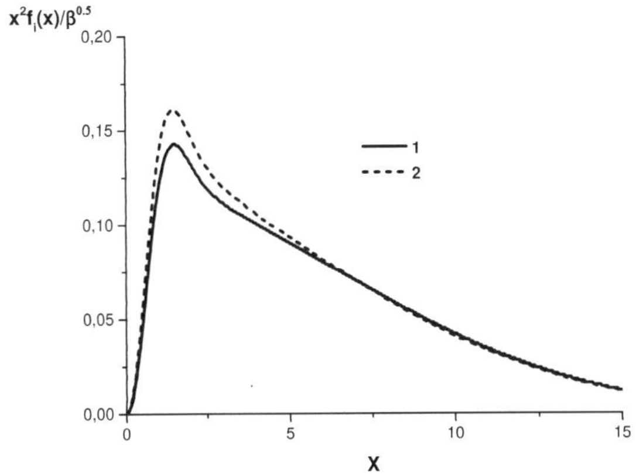

Figure 3. Distribution functions for He+ ions for ρ = 0, $\frac{\alpha_0}{\beta}$ = 0.01, *Ta* = 300 K; (*1*) calculation disregarding the velocity dependence of the charge-exchange cross section; (*2*) calculation taking this dependence into account.

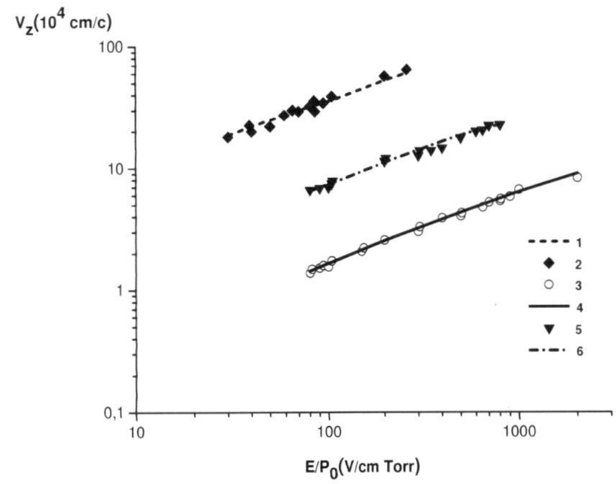

Figure 4. Comparison of the dependence of the drift velocity of He^+ ions in He, Hg^+ ions in Hg vapor, and Ar^+ in Ar in strong fields, calculated from distribution function (5), with experimental data, $P_0 = P \times 273.16/\ T_a$ (*K*); (*1*) calculations based on the developed theory for He^+ in He; (*2*) experimental data [34]; (*3*) experimental data [35]; (*4*) calculations based on the developed theory for Hg^+ in Hg vapor; (*5*) experimental data [34, 36]; and (*6*) calculations based on the developed theory for Ar^+ in Ar.

Figure 3 shows the results of comparison of the IDFs for ρ = 0 and $\frac{\beta}{\alpha_0}$ = 100 for He^+ ions at room atomic temperature taking into account the dependence of the charge-exchange cross section on the velocity and disregarding this dependence. It can be seen that these functions are slightly different in the region of the low temperature extremum in view of the difference in the cross sections at such moderate velocities.

To verify our results, we calculated the drift velocity of the ions in the parent inert gas plasma and in mercury vapor using the IDF obtained theoretically. The experimental data and the results of calculation of distribution function (5) for the drift velocity of He^+ ions in He, Hg^+ in Hg vapor, and Ar^+ in Ar [34–36] are compared in Fig. 4. In the calculation of this velocity, we assumed that expression (6) obtained for the concentration is correct, strictly speaking, only for $\frac{\alpha_0}{\beta} \to 0$. Therefore, we have used the normalization factor $\int_0^\infty F_{i\varepsilon}(\varepsilon_i)d\varepsilon_i$ in these calculations and everywhere below. The calculations were performed taking into account the dependence of the charge-exchange cross section on the velocity. The data on the cross sections for He+ were borrowed from [21, 22], for Hg+ from [22], and for Ar^+ from [36]. The references to experimental data are given in the figure. It can be seen that the correlation between the measured and calculated drift velocities is satisfactory, which confirms the correctness of the defined formulas once again. It should be noted that in spite of the fact that expressions (6) and (6b) were derived under the assumption that inequality (2b) holds, the results of comparison lead to the conclusion that the given theory can be used if the parameter appearing on the left-hand side of inequality (2b) exceeds 3. Figure 5 shows the dependence of the resultant ion distribution function (5) on the azimuthal angle θ for $\frac{\alpha_0}{\beta} = 0.01$ (for example, in He this corresponds to $\frac{E}{P} \approx 300 \frac{V}{cmTorr}$) for different relative velocities $x = v_i\sqrt{\beta} = 0.1$ and 5 in the absence (ε = 0) and presence (ε = 1) of the ambipolar field. It is interesting to note that for $x = 0.1$, the distribution is close to isotropic in spite of the strong field and the presence of the ambipolar field. This is due to the fact that the group of ions with velocities $x \ll 1$ is formed by ions that have been generated as a result of charge exchange, that have the Maxwellian (and isotropic) distribution, and have not been accelerated by the electric field. With increasing relative velocity, the angular velocity distribution for ions becomes more and more extended in the direction of the field (see Fig. 5). The results represented in this figure also show that in the presence of the ambipolar field, the peak of the angular distribution function deviates from the discharge axis (θ = 0); for example, for parameter ε = 1 (ambipolar field is equal to the axial field), the peak corresponds to angle $\theta = \frac{\pi}{4}$.

As noted above, apart from theoretical calculation of the IDF in the parent gas in strong fields, we also measured it in mercury vapor at a low pressure (of the order of 10^{-3} Torr) for the parameter $\frac{E}{P} \approx 400 \frac{V}{cm\,Torr}$ using a planar one-sided probe [27, 28]. The second derivative of the probe current was recorded using the demodulation method [28]. It is well known that the instrument function in this method has the form [30]:

$$A(z) = \frac{8}{\pi}\int_{\frac{|z|}{2\sqrt{2}}}^{1}\left[\frac{u^2 - \frac{z^2}{8}}{u}(1-u)\right]^{0.5} du \quad for\ |z| < 2\sqrt{2}$$

$$A(z) = 0 \quad for\ |z| > 2\sqrt{2}, \tag{12}$$

where $z = \frac{\sqrt{2}(\varepsilon\prime - \varepsilon)}{\Delta\varepsilon}$; and the amplitude of the differentiating signal is Δε.

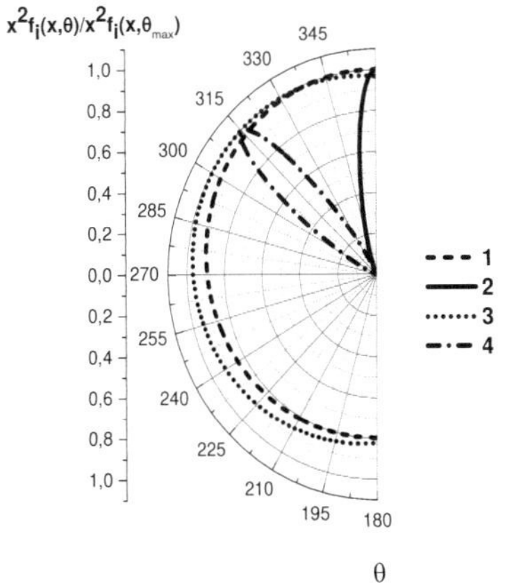

Figure 5. Angular velocity distribution function for ions at $\frac{\alpha}{\beta} = 0.01$; (*1*) $x = 0.1$, ε = 0; (*2*) $x = 5$, ε = 0; (*3*) $x = 0.1$, ε = 1; and (*4*) $x = 5$, ε = 1.

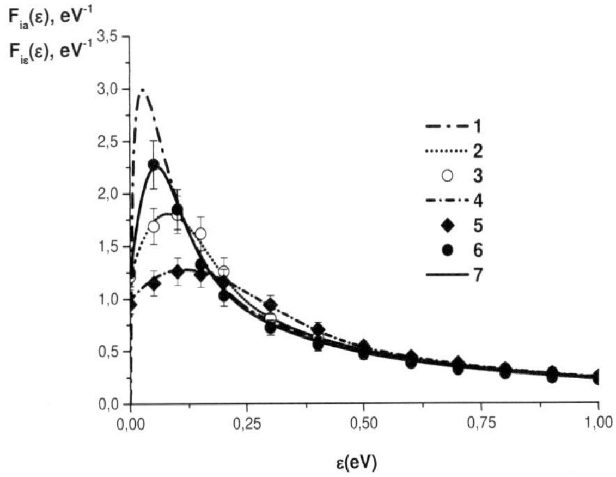

Figure 6. Comparison of the IDF (normalized to unity) for Hg^+ ions in Hg vapor, calculated by formula (6) and measured using the method developed in [29] with a planar one-sided probe for different values of the differentiating signal of the probe method: (*1*) $F_{i\varepsilon}(\varepsilon)$; (*2*) calculation of $F_{ia}(\varepsilon), \Delta\varepsilon = 0.1$; (*3*) experimental $F_{i\varepsilon}(\varepsilon), \Delta\varepsilon = 0.1$; (*4*) calculated $F_{i\varepsilon}(\varepsilon), \Delta\varepsilon = 0.2$; (*5*) experimental $F_{i\varepsilon}(\varepsilon), \Delta\varepsilon = 0.2$; (*6*) calculated $F_{i\varepsilon}(\varepsilon), \Delta\varepsilon = 0.05$; (*7*) experimental $F_{i\varepsilon}(\varepsilon), \Delta\varepsilon = 0.05$. Current density j=100 *mA/cm*^{2}, pressure P=10^{-3} *Torr*, parameter E/P = 400 V/(cm Torr), and atomic temperature T_a = 410 K.

Figure 6 shows the experimental data and the results of corresponding calculations of the convolution of the IDF normalized to unity (6b) and the above instrument function $A(z)$ for various differentiating signals. It can be seen that the measured and calculated dependences are in good agreement.

Figures 7 and 8 show the calculated and measured energy dependences of the coefficients of the IDF expansion in Legendre polynomials of degrees 0– 3 and 4–6, respectively, for the conditions of Fig. 6. It should be noted that the calculations were performed without fitting parameters using the measured value of parameter $\frac{E}{p} = 400 \frac{V}{cm\,Torr}$ and the atomic temperature (equal to 410 K) determined earlier from the coincidence of the calculated and experimental energy distribution functions. It can be seen that good agreements with the results of calculations take place.

Figures 9–11 show the angular dependences for various energies of the calculated IDF normalized to unity, of the calculated sum of the first seven terms of the IDF expansion in the Legendre polynomials, and the same sum, but using the experimentally determined expansion coefficients for the Hg^+ ions under the same conditions as before. It can be seen that for the smallest anisotropy of the distribution function, which corresponds to the minimal energies at 0.05 eV, the coincidence of all three function is very good. Bearing in mind that anisotropy decreases with energy, we can state that for energies lower than 0.05 eV, the coincidence will also be good. With increasing energy (i.e., with increasing anisotropy), the discrepancy is observed between the distribution functions calculated from the first seven Legendre polynomials and the exact function, which is quite natural for such a high value of $\frac{E}{p} = 400\ \frac{V}{cm\,Torr}$.

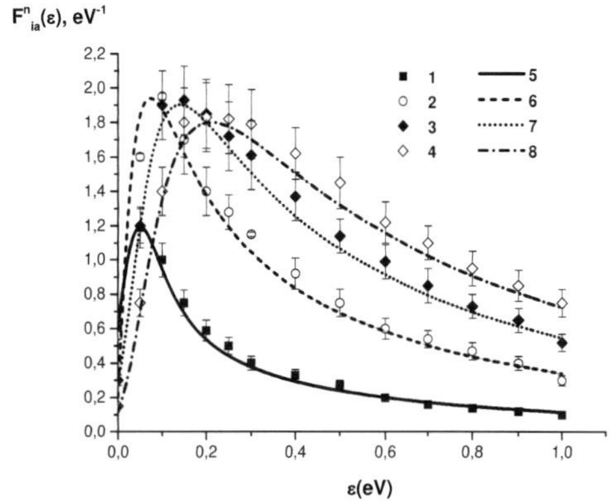

Figure 7. Dependence of the first four Legendre coefficients in the IDF expansion over the directions of their motion under the same discharge conditions as in Fig. 6. The instrument function width is Δε = 0.05 eV; experiment: (*1*) $n = 0$, (*2*) $n = 1$, (*3*) $n = 2$, (*4*) $n = 3$; theory: (*5*) $n = 0$, (*6*) $n = 1$, (*7*) $n = 2$, and (*8*) $n = 3$.

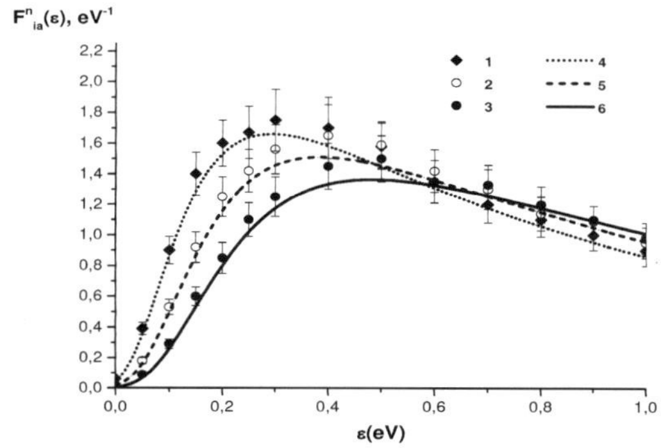

Figure 8. The same as in Fig. 7 for Legendre coefficients 4–6: experiment: (*1*) $n = 4$, (*2*) $n = 5$, (*3*) $n = 6$; theory: (*4*) $n = 4$, (*5*) $n = 5$, and (*6*) $n = 6$.

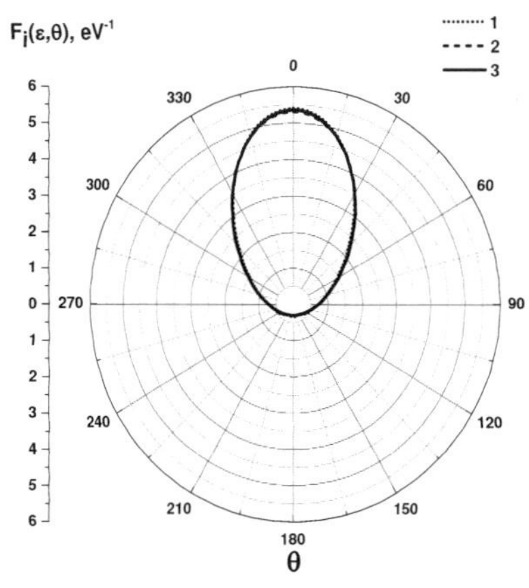

Figure 9. Comparison of angular dependence of the calculated IDF (formula (6)), calculated sum of the first seven terms of the DF expansion in Legendre polynomials, and the same sum determined from measurements for ion energy ε = 0.05 eV; the instrument function width in the probe method is Δε = 0.05 eV. The discharge conditions are the same as in Fig. 6: (*1*) experimental reconstruction using the coefficients of the DF expansion in Legendre polynomials, $n = 0, \ldots, 6$; (*2*) calculated sum of the first seven terms of the DF expansion in Legendre polynomials; and (*3*) calculated DF.

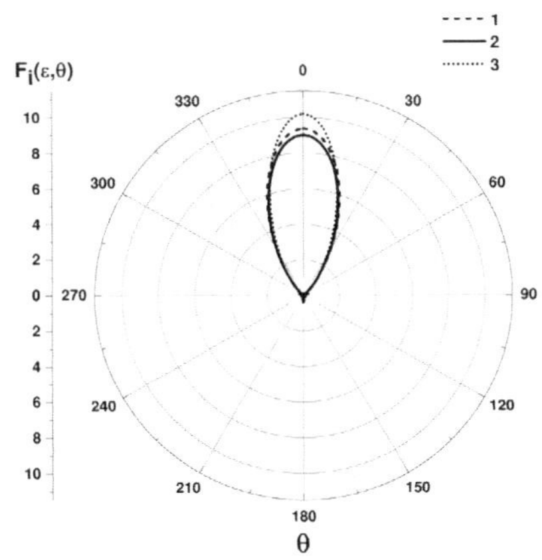

Figure 10. The same as in Fig. 9 for ε = 0.2 eV.

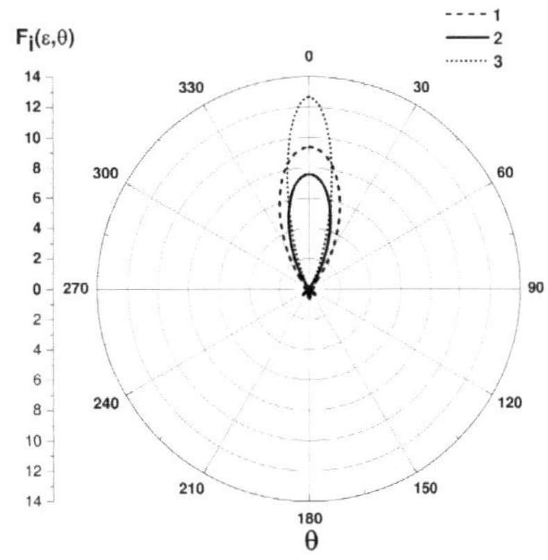

Figure 11. The same as in Fig. 9 for ε = 0.5 eV.

It should be noted that for the plasma of inert gases, the simultaneous fulfillment of inequalities (2b) and (4a) required for the validity of the theory formulated above is ruled out in view of a low ratio $\frac{E}{P}$ for a small role of ambipolar diffusion, a high resonant charge exchange cross section, and a relatively small cross section of elastic scattering from a neutral atom [24]. Thus, to solve the problem of determination of the ion VDF in a glow discharge of inert gases in strong fields outside the discharge symmetry axis, the ambipolar field should be taken into account.

5. Conclusions

The velocity distribution of ions during their motion in the parent gas under the condition that the velocity acquired by an ion over the mean free path is higher than the average velocity of thermal motion of atoms is described by a function differing substantially from the equilibrium Maxwellian distribution. The average and the square mean velocities of ions in strong fields are determined by the energy acquired by an ion over its mean free path, while the most probable velocity is close to the most probable velocity of neutral atoms. For this reason, the representation of the velocity IDF as an equilibrium function with a temperature differing from the temperature of neutral atoms is a quite rough approximation.

The presence of the ambipolar field comparable with the axial field considerably distorts the ion velocity distribution function.

It was measured the energy dependence of the Legendre components of IDF F_{ia}^{n} and it was analyzed the accuracy of the measurements. The theoretical and experimental values of F_{ia}^{n} have been compared. It was obtained a good coincidence of experimental and theoretical results. In the positive column of the electric discharge, the concentrations of electrons and ions are almost identical; therefore, comparison of concentrations obtained from the processing of the experimental curve I_U'' is also one of criteria of the validity of the above interpretation. The difference in the concentrations in our study did not exceed 10%, which is a satisfactory results on account of the error in probe measurements.

Good coincidence of the calculated and measured IDFs leads to the conclusion that the DF of the Hg+ ion under the given conditions is formed as a result of resonant charge exchange, and elastic collisions do not affect its form. It should be noted that, for example, elastic collisions in He plasma in strong fields should be taken into account in the calculation of the DF for the He^{+} ions [37].

Thus, the probe method developed here makes it possible to reliably determine angular harmonics of IDF F_{ia}^{n} , which considerably expands its potentialities. It was measured the energy dependence of the harmonics of the ion distribution function $F_{ia}^{0} - F_{ia}^{6}$. New possibilities of application of the probe method in determining IDF and the convective velocity of ions in a plasma have been illustrated.

References

[1] Ivanov Y and Polak L 1975 *Plasma Chemistry* Ed. by B. M. Smirnov (Atomizdat, Moscow) **2** 161–198
[2] Sena L 1946 *Zh. Eksp. Teor. Fiz.* **6** 34
[3] Kagan Y and Perel V 1954 *Dokl. Akad. Nauk SSSR* **98** 575
[4] Kagan Y and Perel V 1955 *Zh. Eksp. Teor. Fiz.* **29** 884
[5] Smirnov B 1966 *Sov. Phys. Tech. Phys.* **11** 520
[6] Perel V 1957 *Sov. Phys. JETP* **5** 436
[7] Fok V 1948 *Zh. Eksp. Teor. Fiz.* **18** 1048
[8] Golant V, Zhilinskii A and Sakharov S 1977 *Foundations of Plasma Physics* (Atomizdat, Moscow) 384
[9] Ender A and Ender I 2010 *Tech. Phys* **55** 166
[10] Ender A and Ender I 2010 *Tech. Phys.* **55** 1400
[11] Else D, Kompaneets R and Vladimirov S 2009 *Phys. Plasmas* **16** 062106
[12] Viehland L and Mason E 1975 *Ann. Phys.* **91** 499
[13] White R, Robson R, Dujko S, Nicoletopoulos P and Li B 2009 *J. Phys. D: Appl. Phys.* **42**, 194001
[14] Bhatnagar P, Gross E and Krook M 1954 *Phys. Rev.* **94** 511
[15] Frish S and Kagan Y 1947 *Zh. Eksp. Teor. Fiz.* **17** 577
[16] Israel D, Riemann K and Tsendin L 2006 *J. Appl. Phys.* **99** 093303
[17] O'Connell D, Zorat A, Ellingboe A and Turner M 2007 *Phys. Plasmas* **14** 103510
[18] Chen W, Zhu X, Zhang S and Pu Y 2009 *Appl. Phys. Lett.* **94** 211503
[19] Mc Daniel E and Mason E 1973 *The Mobility and Diffusion of Ions in Gases* (Wiley, New York) 373
[20] Nikitin E and Smirnov B 1978 *Sov. Phys. Usp.* **21** 95
[21] Maiorov S 2007 Calculation of resonant charge exchange cross-sections for neon, argon, krypton, xenon, rubidium, caesium, and mercury ions *Proceedings of the 34th International Conference on Plasma Physics and Controlled Fusion* (Zvenigorod) **1** 86
[22] Smirnov B 2001 *Phys. Usp.* **44** 221
[23] Dalgarno A and Bates D 1962 *Atomic and Molecular Processes* (New York) 919
[24] Raizer Y 1991 *Gas Discharge Physics* (Springer, Berlin) 536
[25] Demidov V, Kolokolov N, and Kudryavtsev A 1996 *Probe Methods Research of Low Temperature Plasma* (Energoatomizdat, Moscow) 235
[26] Godyak V and Demidov V 2011 *J. Phys. D: Appl. Phys.* **44** 233001
[27] Lapshin V and Mustafaev A 1989 *Sov. Phys. Tech. Phys.* **34** 150
[28] Mustafaev A 2001 *Tech. Phys.* **46** 472
[29] Mustafaev A, Mezentsev A and Simonov V 1984 *Sov. Phys. Tech. Phys.* **29** 1263
[30] Volkova L, Demidov V, Kolokolov N and Kralkina E 1984 *Teplofiz. Vys. Temp.* **22** 757
[31] Mott-Smith H and Langmuir I 1926 *Phys. Rev.* **28** 727
[32] Ivanov Y, Lebedev Y and Polak L 1981 *Methods of Contact Diagnostics in Non-Equilibrium Plasma Chemistry* (Nauka, Moscow) 144
[33] Berger E and Heisen A 1975 *J. Phys. D: Appl. Phys.* **8** 629
[34] Hornbeck A 1951 *Phys. Rev.* **84** 615
[35] Kovart F 1964 *Phys. Rev. A* **133** 681
[36] Hegerbergt R, Elford M and Skulleruds H 1982 *J. Phys. B: At. Mol. Phys.* **15** 797
[37] Barata J and Conde C 2010 *Nucl. Instrum. Methods Phys. Res. A.* **619** 21

RF plasma probe diagnostics: a method for eliminating measurement errors for Langmuir probes with bare protective shields

V A Riaby[1], P E Masherov[1], V P Savinov[2], and V G Yakunin[2]

[1] Research Institute of Applied Mechanics and Electrodynamics of the Moscow Aviation Institute (State Technical University), 5 Leningrad Road, Moscow 125080, Russia

[2] Moscow State University, Physics Department, Moscow 119991, Russia

E-mail: riaby2001@yahoo.com

Abstract. The new DC arc T-plasmatron of long service life [1] is studied. The well known method of the electric field strength measurements in a stabilized arc channel [2] was applied in a modified form as a consequence of the specific form of the presumably diffuse anode spot attached to a gas vortex on the external surface of the anode unit. The electrical field strength was determined assuming that the potential drop across the diffuse anode spot in the new plasmatron was small. This gave the mean argon plasma conductivity: $\sigma \leq 118$ $Ohm^{-1}cm^{-1}$ for arc currents $I \leq 180$ A which agreed with the independent experiment [2] affirming the correctness of the above assumption. Analysis of the known experimental and theoretic data on atmospheric argon plasma conductivity resulted in the selection of R.S.Devoto's theoretic dependence $\sigma(T)$ [3] as the most reliable one for T=8000...20000 K at P= 1 atm that allowed the evaluation of the mean argon plasma temperature at the exit of the plasmatron: $T \leq 19500$ K.

1. Introduction

The aim of the present work was to identify the reason for the discrepancies in plasma-probe measurements that were obtained by two cylindrical probes in one common cross-section. One probe was straight and radially movable, and the other was L-shaped and shifted from its center so that it could revolve around its own longitudinal axis to register the same radial plasma-parameter distributions as the straight probe. Plasma-parameter measurements were conducted in the inductive gas-discharge chamber (GDC) of a radio-frequency (RF) ion-thruster model using a planar antenna coil with a ferrite core and operating at the driving frequency f = 2 MHz. The thruster's ion-extracting grate (IEG) was replaced by a meshwork in the GDC pumping sleeve to provide xenon plasma pressure p = 2 mTorr at the flow rate of q = 2 sccm. Plasma properties were measured using both probes in the middle cross-section of the thruster's discharge space at a distance of 33 mm from the quartz window separating the antenna coil from the plasma. This research is necessary for effective thruster's IEG development. Both measuring probes comprised reference probes with large collecting surfaces to protect the probe's volt-ampere characteristics (VACs) against RF distortions. The probe leads were shielded by bare metal tubes to eliminate RF interferences in the probe circuits as has been done in previous research [1] dealing with microwave plasma. In the present work, the results of the two probe measurements differed quite noticeably, especially in two common points, $r = 0$ (plasma space axis) and $r = 60$ mm (radial position), located 13 mm from the vacuum-chamber wall. The present work aimed to find the reason for this situation, a way to eliminate the measurement errors and to analyze its physical nature.

2. Experimental facility and plasma diagnostics arrangement

The schematic diagram of the ion thruster model was presented in work [2] where plasma parameters were measured using both types of probes, although that work was dedicated to the results obtained using only the straight probe. Figure 1 shows the draft of the thruster model.

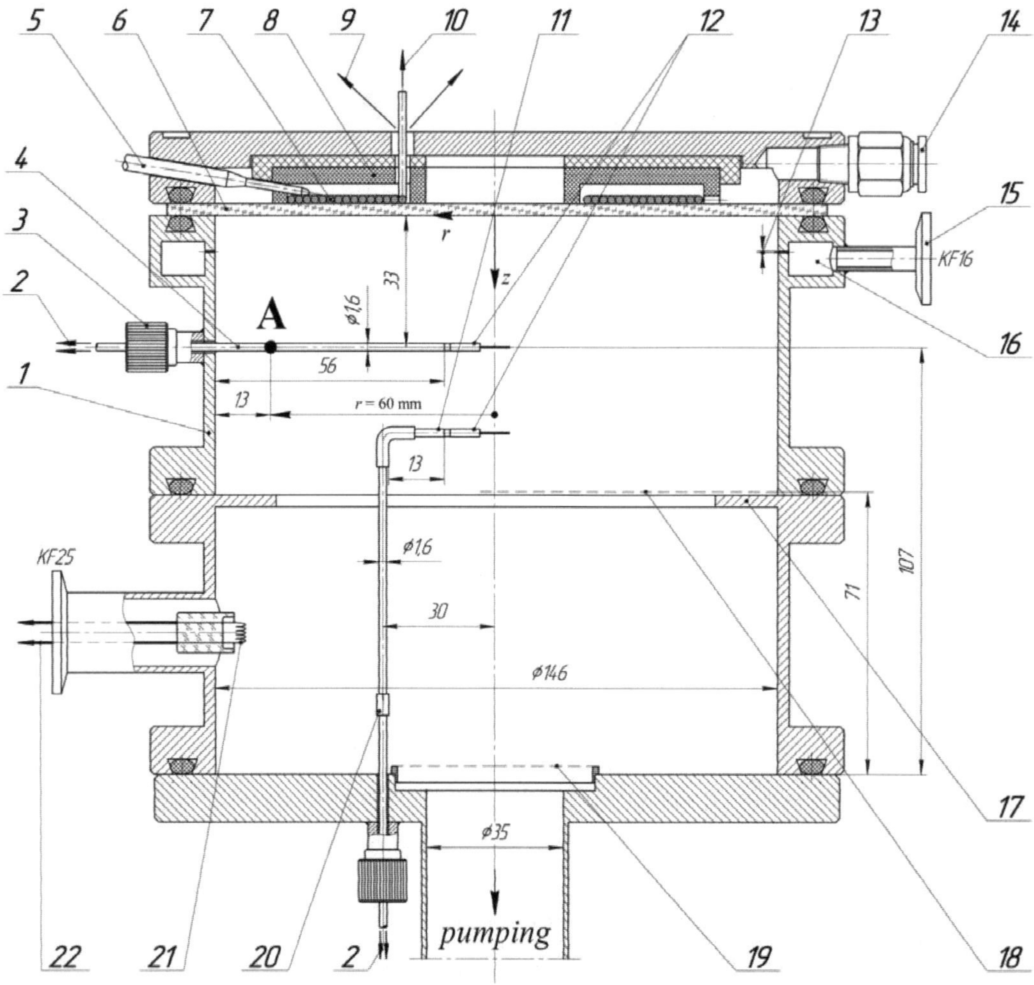

Figure 1. Drawing of the ion thruster model. 1. Vacuum chamber. 2. To the VGPS-12 probe station. 3. Movable UltraTorr vacuum seal. 4. Straight cylindrical probe. 5. Temperature-sensitive gauge. 6. Quartz window. 7. Planar antenna coil. 8. Ferrite core. 9. Cooling-air output. 10. To the matching network. 11. L-probe. 12. Reference probes. 13. One of eight xenon feeding holes, 0.4 mm in diameter. 14. "One-touch" sleeve for cooling-air input. 15. Xenon feeding KF16 fitting. 16. Xenon feeding circular collector. 17. IEG-mounting shoulder. 18. Lowest position of the L-probe. 19. Meshwork. 20. Metal sleeve connecting two parts of the L-probe shield. 21. Tungsten filament of an opened halogen lamp. 22. To the electronic transformer of the halogen lamp.

The general view of this draft was presented in previous research [3] that studied a by-wall plane probe simulator. Following engineering traditions, this draft shows the details of two different cylindrical probe designs, including expanded dimensional data, which is very important for evaluations of the discharge plasma perturbations caused by both types of probes.

In the present work, the probe diagnostics were arranged as follows. The radially movable, straight cylindrical probe 4 could be moved in the radial position range r = 0–60 mm, and at r = 60 mm (special point A), the probe shield was deepened by 4 mm into the body of the vacuum fitting 3. Therefore, the range of its shield lengths was l_{sh} = 0–56 mm. The L-probe 11 was inserted into the GDC through the port located at the bottom of the vessel, 30 mm from the axis of the discharge space, and could move in the range of longitudinal positions z = 33–69 mm, during which its vertical shield length varied in the range of l_{sh} = 71–107 mm with 13 mm of the shield length of its shoulder, providing a range of full shield length equal to l_{sh} = 84–120 mm. These measurements will be used below to analyze the probe measurement results.

Probe diagnostics of the xenon plasma were conducted using the automated probe station VGPS-12, which is described on the Plasma Sensors Company's web site [4] and briefly characterized in a review [5]. In this probe system, the objectivity of measurements was assured by using the most advanced experimental physics achievements, enhanced by the VGPS-12 creators' multiple years of experience. Analysis of the quality and objectivity of the VGPS measurements showed that it registered electron temperature T_e and electron concentration n_e within an error margin of approximately ± 10% due to a wide dynamic range of EEPF registration (approximately 60 dB and more), resulting in highly precise EEDF determination [6]. The plasma potentials, floating V_f and plasma space V_s, determining electron saturation current densities j_{es}, were measured with much less errors.

The cylindrical probes used in the present work were made of tungsten filament 0.15 mm in diameter. Their length was 10 mm, which ensured negligible local plasma perturbations caused by charged particle recombination on the probe holder, which was 1.6 mm in diameter [7]. Figure 2 shows a draft of the straight cylindrical probe.

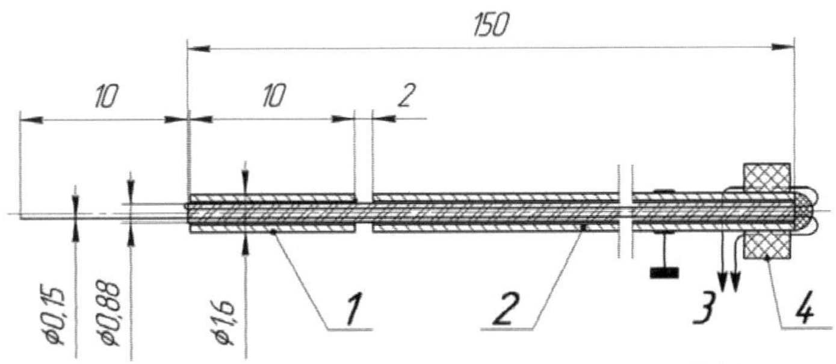

Figure 2. Straight cylindrical probe. 1. Reference probe. 2. Protective shield. 3. To the VGPS-12 probe station. 4. Termination unit for probe leads.

The tip of the L-shaped probe and its body, including the reference probe and the protective shield, were made similarly, with a rectangular bend that had a shoulder with a radial size of 30 mm from the vertical axis of its probe holder to the middle of the probe tip, as shown in figure 1.

3. Results of probe measurements

Probe measurements were conducted at z = 33 mm in the middle cross-section of the gas-discharge space, where the straight cylindrical probe moved radially in the range r = 0–60 mm, and the L-probe revolved, passing three radial points: r = 0, 30, and 60 mm. The two probes had two common measurement points: r = 0 and r = 60 mm. In the axial position, both probe shields interacted with surrounding plasma, while at the point r = 60 mm, the straight probe's shield was absent (l_{sh1} = 0), and the L-probe's shield had a maximal length of l_{sh2} = 120 mm. Figures 3–6 show the results of the T_e, n_e,

V_s, and j_{es} radial-distribution measurements obtained by both probes for the incident RF generator (RFG) powers P_{in} = 50–200 W. The straight probe's data are represented by solid lines and the L-probe's data by dashed lines.

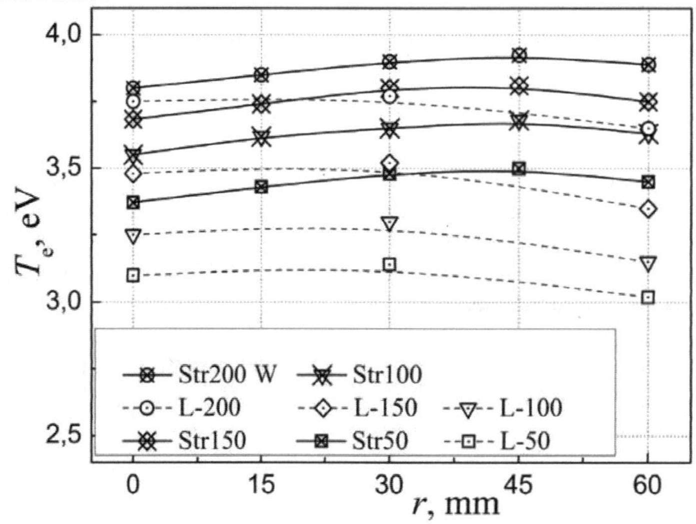

Figure 3. Radial distributions of T_e for two probe types at P_{in} = 50–200 W. Str = straight probe; L = L-probe.

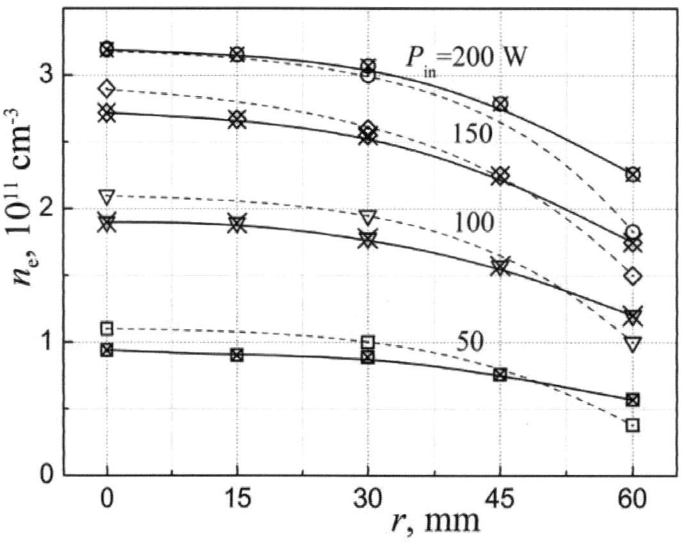

Figure 4. Radial distributions of n_e for two probe types at P_{in} = 50–200 W.

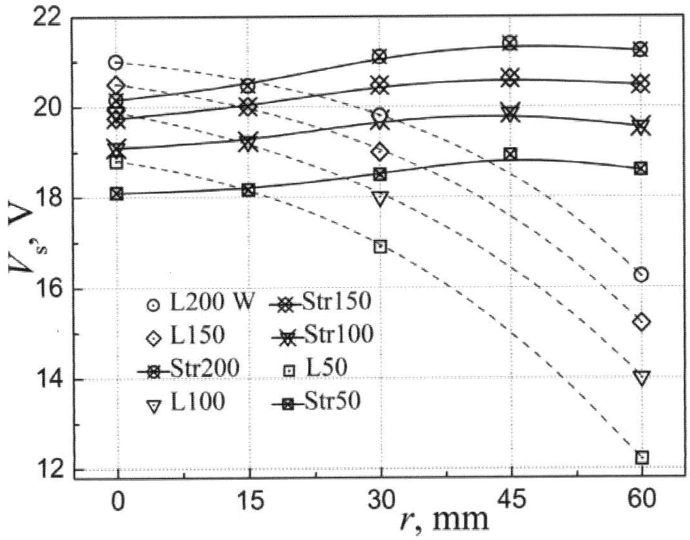

Figure 5. Radial distributions of V_s for two probe types at P_{in} = 50–200 W. Str = straight probe; L = L-probe.

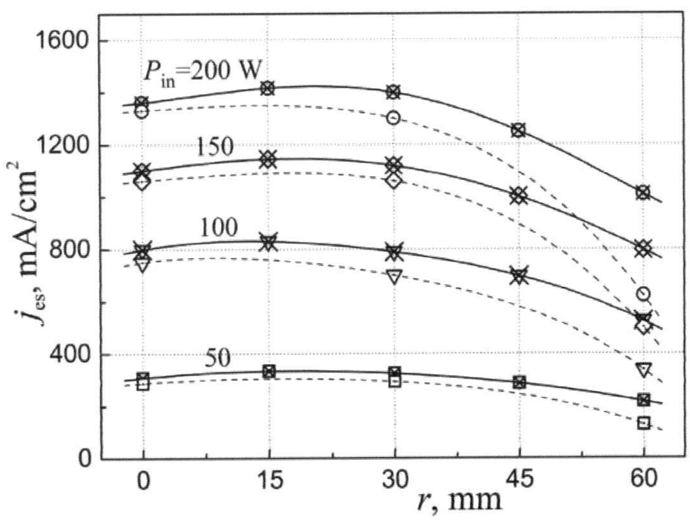

Figure 6. Radial distributions of j_{es} for two probe types at P_{in} = 50–200 W.

The readouts of the two probes differed quite noticeably, although they registered plasma parameters in the same cross-section, and they had two coincident points at which their readouts differed quite noticeably, especially in the second position, r = 60 mm, which was beside the GDC wall. Their general deviations were more substantial for plasma-space potentials V_s (figure 5) and for electron temperatures T_e (figure 3). As mentioned previously, due to the negligibility of local plasma distortions caused by the tips of both probes [7] and the use of protective reference probes and shields, we suspected that the probe-measurement discrepancies obtained in the present work were caused by some disturbances of the gas-discharge space structure. Possible explanations for these discrepancies include different probe designs, different ranges in shield-length variations, and their different locations. However, the question remained, what is the main reason for these measurement peculiarities?

4. The reason for measurement data discrepancies

In search of the answer to this question, we tried to analyze measured plasma EEDFs and their deviations from the ideal Maxwellian functions. This task was natural because the control program of the VGPS-12 probe station was based on the Druyvesteyn method, according to which, as the first step of the probe diagnostics, plasma EEDFs were registered very accurately and without any limitations of the EEDF forms. To reach a clear solution, we used the quantitative method for EEDF evaluation proposed in our previous work [8], because a qualitative comparison of EEDF appearances could not yield definitive conclusions. According to this method, the measured electron saturation current densities j_{es} are compared with theoretical Maxwellian ones using the ratios j_{es}/j_{esM}, where j_{esM} is the electron saturation's current density for the ideal isotropic Maxwellian plasmas calculated by the following formula:

$$j_{esM} = (1/4)en_e(8eT_e/\pi m_e)^{1/2} \quad (1)$$

using the measured plasma parameters n_e and T_e, where e is the elementary electron charge, T_e is expressed in V, and m_e is the electron mass.

To simplify the forms of the symbols, we assumed that the ratio j_{es}/j_{esM} was denoted by R_M, the straight probe by 1, and the L-probe by 2. The radial distributions $R_M(r)$ were determined for both probes. Figure 7 shows the results of these calculations for the straight probe at P_{in} = 50–200 W.

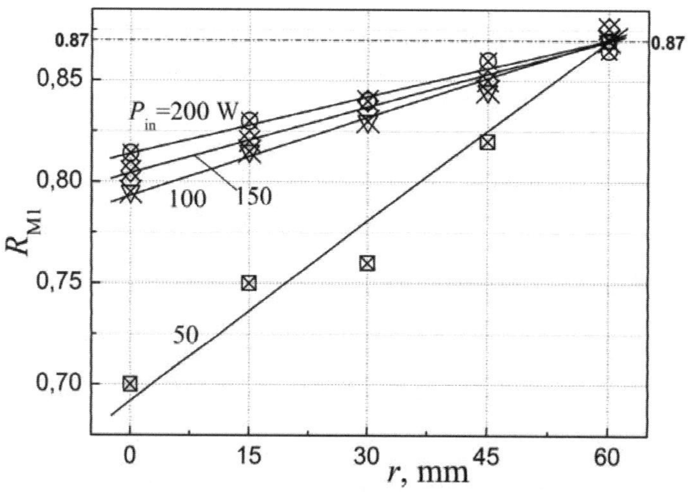

Figure 7. Radial R_{M1} distributions for the straight probe at P_{in} = 50–200 W.

These results were interesting and unexpected, showing that for the straight probe, EEDFs and their deviations from the ideal Maxwellian functions varied linearly with the length of the shield l_{sh1} from the maximal spread R_{M1} = 0.7–0.82 at r = 0 and l_{sh1} = 56 mm. This spread linearly converged and its mean level also linearly rose with l_{sh1} to the single point $R_{M1} \approx 0.87$ at r = 60 mm that remained approximately constant for all levels of incident RFG power P_{in} at the point where the shield was absent, l_{sh1} = 0. Therefore, possible plasma perturbations around the straight probe were proportional to the length l_{sh1} of its shield, achieving minimal perturbation level at the special point A (l_{sh1} = 0, r = 60 mm – see figure 1), which can be considered as the point at which the probes of both types having determined the full set of plasma parameters enabled quantitative comparison of their values. Figure 8 shows the analogous data for the L-probe.

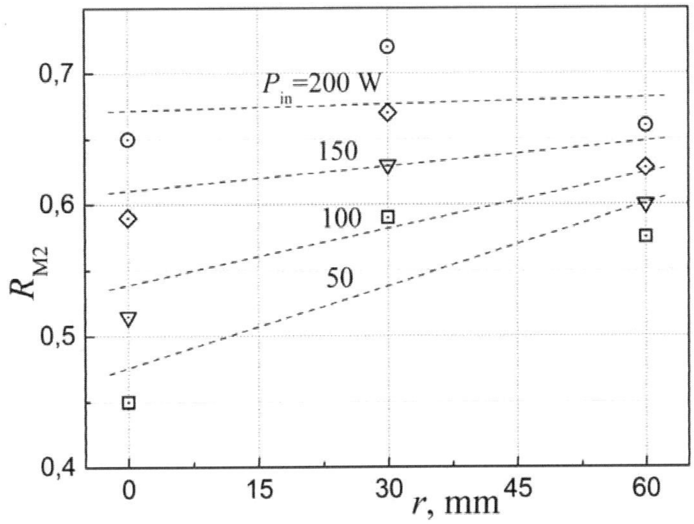

Figire 8. Radial R_{M2} distributions for the L-probe at P_{in} = 50–200 W.

The radial distributions $R_{M2}(r)$ for the L-probe showed that its R_{M2} ratios varied in lower ranges: 0.45–0.65 at $r = 0$, 0.6–0.72 at $r = 30$ mm, and 0.57–0.65 at $r = 60$ mm. In addition, their linear approximations demonstrated that the general EEDF perturbations for the L-probe weakly decreased toward the vacuum chamber wall that corresponded to a slight rise of $R_{M2}(r)$ approximating lines. At the special point A ($l_{sh1} = 0$, $r = 60$ mm), the EEDFs for the plasma around the L-probe deviated from the Maxwell function much more than the 13% deviations for the plasma around the straight probe corresponding to $R_{M1} \approx 0.87$. This fact means that at this point, the plasma-shield interactions for the straight probe were rather weak, but for the L-probe, they were much more active.

Therefore, measurement data for the L-probe 2 at this special point A ($l_{sh1} = 0$, $r = 60$ mm) can be quantitatively compared with the practically distortion-less data for the straight probe 1 to determine the actual plasma perturbations caused by the L-probe's long shield. This was done by determination of the ratios T_{e2}/T_{e1}, n_{e2}/n_{e1}, V_{s2}/V_{s1}, and j_{es2}/j_{es1} using the special point's data for both probes from figures 3–8. Figure 9 presents thus obtained experimental results in the form of $(x_2/x_1)(R_{M2})$ dependences for the L-probe at various levels of RFG incident power P_{in} united with $R_{M1} \approx 0.87$ corresponding to $(x_2/x_1) = 1$ for the straight probe.

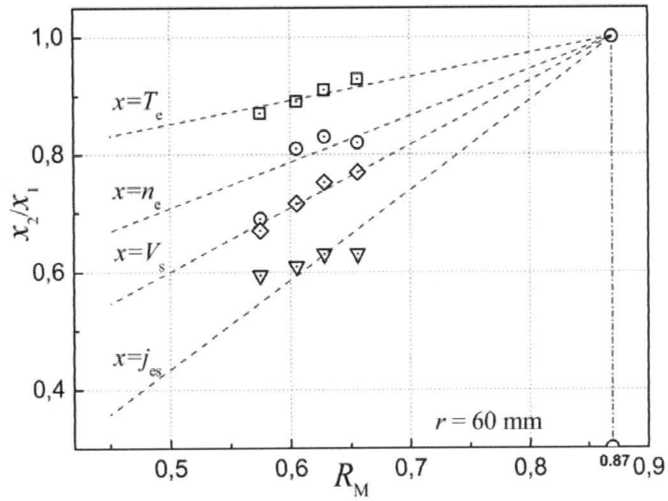

Figure 9. Linearly approximated functions $(x_2/x_1)(R_M)$ in the range of R_M = 0.45–0.87.

These $(x_2/x_1)(R_M)$ dependences together with the point $[(x_2/x_1) = 1, R_M \approx 0.87]$ were linearly approximated to the minimal level of $R_{M2} \sim 0.45$ that is shown in figure 8. This is a very important universal dependence that quantitatively characterizes the influence of EEDF perturbations on the measured probe parameters and allows transforming the $R_M(r)$ functions from figures 7 and 8 into $(x_2/x_1)(r)$ dependences for definite plasma properties that can be denoted as $(x_D/x_{UD})(r)$, where x_D is a disturbed plasma parameter and x_{UD} is an undisturbed one. These ratios can be used to reform the disturbed measurement results $x_D(r)$ into the corrected functions $x_{UD}(r) = x_{Corr}(r)$.

By comparing measurement results for the L-probe with the data for the straight probe at the special point A ($l_{sh1} = 0$, $r = 60$ mm), we quantitatively evaluated the L-probe measurement errors that were caused by the EEDF distortions for the plasma around its probe tip. Analysis of figure 9 showed that at the special point A, the L-shaped probe lowered T_e by up to 13%, n_e by ≤ 28%, V_s by ≤ 33%, and j_{es} by ≤ 40%. These data resulted from parallel use of cylindrical probes of two different designs inserted into the vacuum chamber in two different locations with realization of the aforementioned special point A, which allowed us to determine the L-probe measurement errors caused by the bare shields contacting the RF plasma. These peculiarities in probe diagnostics presented quite unexpected data never before published and discussed in physics literature.

5. A method of RF plasma diagnostics error elimination for probes having bare protective shields

The abovementioned measurement errors for Langmuir probes having bare protective shields in contact with RF plasmas should be impossible when the probe shields are covered by dielectric coatings for in this case above considered phenomenon causing plasma distortions cannot arise. But when probe shields remain externally uninsulated because of technological difficulties, two unprotected probes should be used to refine the local diagnostics of RF plasmas using the method proposed in the present work. This technique can be accomplished using the two stages described below.

I. Preparing probe measurement corrections.

a) One probe, which we denote as probe No. 1, should be moved in the RF discharge space, varying the length of its shield in the range of $l_{sh1} = 0–l_{sh1m}$, where l_{sh1m} is the maximal shield length for probe No. 1. Another probe, called probe No. 2, should have the possibility of reaching the special position A of probe No. 1 at $l_{sh1} = 0$ when $l_{sh2} \neq 0$. From the physical standpoint, it is better to use both probes in the common plasma space cross-section, with l_{sh2m} being not far from l_{sh1m}.

b) After probe measurements of the plasma parameters in the form of disturbed functions $x_{1D}(l_{sh1})$ and $x_{2D}(l_{sh2})$ for $x = T_e$, n_e, V_s, and j_{es} and EEDFs determinations, quantitative evaluations of the EEDF deviations from the Maxwell function can be made as $(j_{es}/j_{esM}) = R_M$, using the measured values of j_{es} and those calculated with Eq. (1). In the case of probe No. 1, when $l_{sh1} = 0$, the maximal R_{M1} value, R_{M1m}, should represent a single upper R_M parameter for all levels of the RFG incident powers P_{in}. These actions should yield the $R_M(l_{sh})$ dependences [or $R_M(r)$ in the case of a circular discharge space] for various P_{in} values.

c) The ratios of the measured disturbed/undisturbed probe parameters (x_{2D}/x_{1UD}) for $x = T_e$, n_e, V_s, and j_{es} and for various P_{in} should be determined as functions of R_{M2} at the special point A at which $l_{sh1} = 0$ with x_{2D} representing the disturbed parameters for probe No. 2 and x_{1UD} representing the undisturbed parameters for probe No. 1.

d) Dependences $(x_{2D}/x_{1UD})(R_{M2})$ should be united with $(x_2/x_1) = 1$ at R_{M1m}, for which the shield influence on probe No. 1 is absent, and their combination should be linearly approximated to obtain the general function $(x_2/x_1)(R_M)$ similar to that in figure 9.

II. Correcting probe measurement results.

a) For a necessary plasma parameter x, the RFG power P_{in} and the probe No. corresponding function $R_M(l_{sh})$ should be transformed into the dependence $(x_D/x_{UD})(l_{sh})$, using the general function $(x_2/x_1)(R_M)$ to exclude the intermediate variable R_M.

b) The initial measurement points $x_D(l_{sh})$ reflecting the plasma disturbances should be divided by corresponding $(x_D/x_{UD})(l_{sh})$ corrective points to obtain the corrected measurement data $(x_{UD})(l_{sh})$ = $(x_{Corr})(l_{sh})$.

As an example, we are showing the correction of the straight probe's measurement data on $T_{e1}(r)$ and $V_{s1}(r)$, from figures 3 and 5, respectively, for the practically useful RFG incident powers P_{in} = 100–200 W. Using dependences $R_{M1}(r)$ from figure 7 for the straight probe and universal functions $(x_2/x_1)(R_M)$ from figure 9, we transform the initial diagnostics results into the correcting functions $(T_{e2}/T_{e1})(r)$ and $(V_{s2}/V_{s1})(r)$, where T_{e2} and V_{s2} were exposed to plasma distortions, and T_{e1} and V_{s1} represented undisturbed parameters. Then, we divide the $T_{e1}(r) = T_{eD}(r)$ and the $V_{s1}(r) = V_{sD}(r)$ points by the $(T_{e2}/T_{e1})(r) = (T_{eD}/T_{eUD})(r)$ and $(V_{s2}/V_{s1})(r) = (V_{sD}/V_{sUD})(r)$ points, respectively, to obtain the functions $T_{eUD}(r) = T_{eCorr}(r)$ and $V_{sUD}(r) = V_{sCorr}(r)$, which are represented in figures 10 and 11 by solid lines, respectively. In these same figures, the initial data on $T_{e1}(r)$ and $V_{s1}(r)$ are shown by dashed curves.

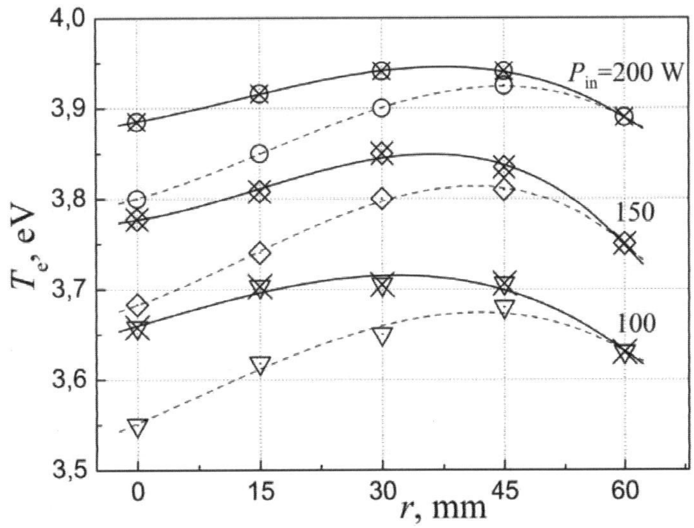

Figure 10. Radial distributions of T_e for the straight probe at P_{in} = 100–200 W. Dashed lines = distorted data; solid lines = corrected curves.

IOP Conf. Series: Journal of Physics: Conf. Series **958** (2018) 012006 doi:10.1088/1742-6596/958/1/012006

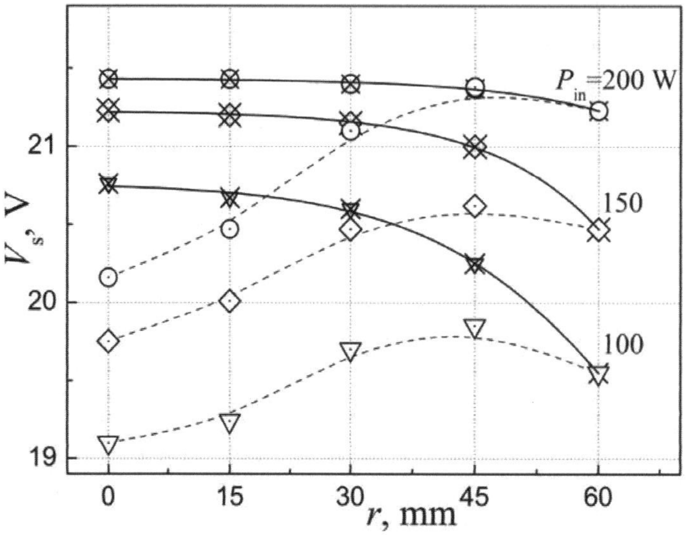

Figure 11. Radial distributions of V_s for the straight probe at P_{in} = 100–200 W. Dashed lines = distorted data; solid lines = corrected curves.

Such corrections can be made for all measured probe parameters. They represent objective plasma properties with some small errors caused by the fact that at the special point (l_{sh1} = 0, r = 60 mm), the reference probe of the straight measuring probe might have excited some residual plasma distortions.

6. Physical nature of the disclosed plasma distortions

The identification of the linear relationships between the EEDF distortions and the length of the straight probe's bare shield indicated that this effect may be related to some particular physics of plasma-shield interactions. To understand their physical nature, we turned to the authors' previous works [9-20] on the behavior of large-scale conducting bodies in contact with plasmas. Some of these works [9-17] found that such bodies, represented by silicon wafers, were exposed to a short-circuited double-probe phenomenon, according to the authors' physics common sense. The correctness of this idea was proved by successful arrangement of the silicon-wafer processings in the RF oxygen plasmas. Direct experimental confirmation of this phenomenon was presented in a report [18] that demonstrated its physical essence and displayed the classical double-probe VAC for a metal body consisting of two parts connected by an ammeter and a variable DC voltage source.

Two more works [19, 20] were dedicated to probe measurements in RF plasmas using bare metal protective shields underneath the floating potential. They showed qualitatively that the short-circuited double-probe phenomenon in such probe shields could lower the plasma-ionization balance, thus decreasing the plasma parameters along with the electrical power loss in the shield. Given this, it was of interest to compare the physical nature of the plasma-shield interactions for the shields underneath the floating potential [19, 20] and for those underneath the ground potential used in the present work.

We analyzed such interactions using the experimentally measured radial distribution of plasma floating potential, $V_f(r)$, which is presented as graph 1 in figure 12.

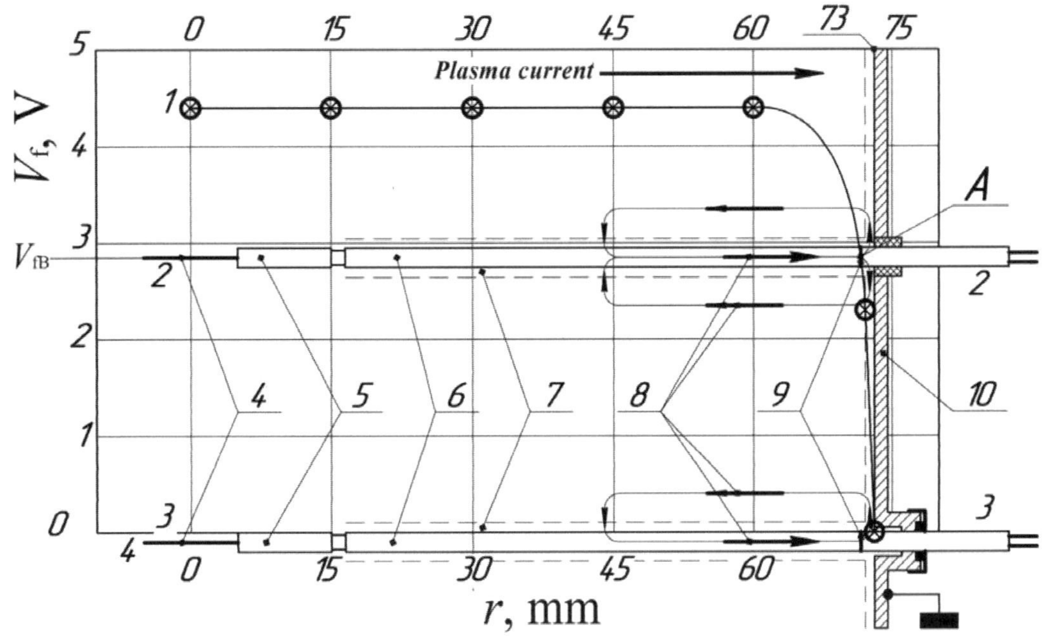

Figure 12. Radial $V_f(r)$ distribution in xenon plasma interacting with two versions of the bare probe shields. 1. Measured radial distribution of plasma floating potential $V_f(r)$. 2. Straight cylindrical probe with protective shield underneath the floating potential. 3. Straight cylindrical probe with grounded protective shield. 4. Measuring probe tips. 5. Reference probes. 6. Protective probe shields. 7. Sheaths of space charge beside probe shields and GDC wall. 8. Short-circuited double-probe currents, I_{SC}. 9. Steady-state lines dividing the collecting surfaces of bare probe shields. 10. Grounded GDC wall. A. Point of $V_f(r)$-probe axis intersection.

The five points on this graph for r = 0–60 mm represent the results of the straight probe measurements: V_f= 4.39 V = Const. for all levels of P_{in}. The point V_f = 2.35 V, at the boundary of the by-wall sheath, was determined using the plane-wall probe simulator and remained constant for the same P_{in} range [3], while V_f = 0 corresponded to r = 73 mm, the internal surface of the grounded vacuum chamber wall.

The straight cylindrical probe 2 shown in figure 12 comprises probe tip 4, reference probe 5, and bare protective shield 6, which is insulated from the GDC wall to operate underneath the floating potential. Previous works [9-20] considered similar conducting bodies, including silicon wafers and large-scale metal bodies immersed in plasma. Experimental results [18] showed that, in this situation, large conducting bodies functioned as asymmetric short-circuited double-probes. This meant that the floating potential V_{fB} (figure 12), which exhibited negligible longitudinal potential variance in the metal shield because of its high conductivity, was established self-consistently on a level determined by the intersection of the shield's axis with curve 1 in point A. This point belongs to line 9, which divides the shield's collecting surface into two unequal parts. According to figure 12, the shield's long, left part (more negative than the plasma) collected positively charged ions, while its short, right part (more positive than the plasma) collected negative electrons, forming a short-circuited current I_{SC}, which flowed along closed path 8 in the metal shield and in the plasma. The collecting surface of such self-established "double-probe" is proportional to the different mobilities of the ions and electrons, making this "probe" asymmetric. The authors' previous works [9-20] showed that, if such a large-scale metal body was oriented along the plasma or discharge current, the plasma branch 8 of the I_{SC} current would flow against this current, thereby reducing it and decreasing the plasma-ionization balance around this body, thereby lowering all plasma parameters. In addition to this effect, the flow I_{SC} in the shield would heat it, increasing the loss of electrical power. These peculiarities represent the physical nature of the plasma distortions caused by large-scale conducting bodies under floating potential in contact with plasmas.

In figure 12, the straight cylindrical probe 3 is shown to comprise the bare probe shield underneath the constant ground potential that was used in the present work. It is evident that the longitudinal variation of the plasma potential in contact with this metal body underneath the zero potential enables positively charged ions to be collected by the probe shield, which is more negative than the plasma. In the present experiment, the abovementioned phenomenon was realized a little bit differently. As did the previous version of the probe, the long, left shield surface collected positively charged ions that entered the shield and dragged the negatively charged electrons to the opposite end. As a result, a voltage decrease across the space-charge sheath that was beside the shield termination in the wall directed electrons into the right part of the probe shield. Due to high electron mobility, this part of the shield was rather small. Therefore, separation line 9 of the collecting surface of this macro-probe was self-consistently fixed to form a short-circuited double-probe current I_{SC}. This current resulted from deformation of the space-charge sheath around the short, right part of probe 3, and it also caused plasma perturbations analogous to those described above.

7. Discussion

In the abovementioned example [1] the probe measurements consisted of two stages. First were measurements with an unshielded ceramic probe that resulted in highly disturbed probe VACs caused by microwave interferences in the probe circuit that did not allow for determination of reasonable plasma parameters. Second were measurements by a probe having a copper protective shield, which resulted in classical probe VACs without microwave interference distortions that allowed determining more-or-less reasonable plasma parameters. The present work proposed a third stage for such an experiment—a refinement of the second stage results—in which all measured probe parameters are uplifted to eliminate the influence of the above considered phenomenon.

The present work showed that it disturbed plasma parameters proportionally to the shield's length. To understand the reality of this situation, we must evaluate the I_{SC} current in the probe shield and in the plasma around it and compare it with the mean current of the inductive discharge $I_P \sim 2$–3 A [8]. Consideration of the experimental probe VACs of the straight probe with a shield 1.6 mm in diameter and up to 56 mm long resulted in evaluating $I_{SC} \sim 28$–56 mA (in the beginning of the probe VAC's ion branches, which should correspond to the shield's short-circuited, double-probe VACs, ion-current densities were about (1–2) × 10^{-4}A/mm^2, with the probe shield's collecting surface ~282 mm^2). This probe was located $z = 33$ mm from the internal surface of quartz window 6 (figure 1). The discharge current in this location should be approximately one order of magnitude lower than the mean discharge current (i.e., about 0.2–0.3 A). Therefore, the I_{SC} around the shield of the straight probe was only one order of magnitude lower than the local discharge current I_P. That is why discharge plasma EEDFs could be rather sensitive to I_{SC}, which was definitely reflected by the results of the present experiment. As for the L-probe, the collecting surface of its protective shield was twice as large, while its distance from the quartz window was much greater. That is why its EEDF distortions were much deeper compared to those of the straight probe.

When discussing the results of the present work, our chief consultant, Prof. V. Godyak, supposed that one of the reasons for the discrepancies in probe diagnostics could be some noticeable RF component of the plasma floating potential V_{fRF}, which, in this experiment, was not measured due to organizational reasons. It is well known that this parameter can cause serious distortions of the probe VACs, resulting in probe measurement errors [5]. While preparing the present experiment, we followed the main features of the V. Godyak's facility [21], using planar antenna coils enhanced by ferrite cores. In this work V_{fRF} was measured in the wide driving frequency range, including our $f = 2$ MHz, for RF power absorbed by plasma $P_P = 100$ W (which, in our facility, corresponded to $P_{in} \approx 120$ W) and for the plasma pressure range $p = (1$–$1000)$ mTorr. According to [21] for $p = 2$ mTorr, this parameter was approximately $V_{fRF} \sim 0.4$ V, which was many times lower than the mean electron temperature $T_e \sim 3.5$ eV measured in the present work (see figure 4). In such situation distortions of probe VACs should be negligible [21]. In general, the V_{fRF} depends on the P_{in} level (in our experiments, it reached 200 W) and on the material properties of the ferrite. Therefore, it would be useful to measure the V_{fRF} in subsequent experiments. However, the present work showed that EEDF distortions from V_{fRF} can hardly depend on the length of the bare probe shield.

The present work will be very useful for all physicists engaged in plasma studies, especially those dealing with RF plasmas, because Langmuir probe diagnostics is a very important technique in plasma-parameter measurement. The present results clarify that the best way to make objective probe measurements of FR plasma parameters is to use protective shields coated with dielectric layers to eliminate the short-circuited double-probe phenomenon addressed in this work. In the case of probes having bare protective shields, it is beneficial to follow the method proposed in the present work.

8. Conclusions

1. Probe studies of the RF plasma were carried out using two cylindrical probes having bare protective shields. They moved in a common cross-section in such a way that in the single **special point A** probe No.1 reached zero shield length with non-zero shield length here for probe No. 2.
2. Measurement results for both probes differed quite noticeably with maximal differences in the **special point A**.
3. The reason for this situation was found using the authors' quantitative method of EEDF evaluations, which demonstrated linear dependence of EEDF deviations from the Maxwell function on the shield length for the probe No. 1.
4. Application of this action to the results of probe measurements in the **special point A** helped to find the relationship between probe diagnostics errors and EEDF distortions and to correct measurement data for all plasma parameters in all probe positions.
5. Following previous conclusion points, a method of measurement error elimination has been proposed in the present work for probes immersed in the RF plasmas and having bare protective shields.
6. Physical nature of plasma distortions disclosed in the present work was the short-circuited double-probe phenomenon in the metal shields.
7. This phenomenon and its influence on the probe measurement results in RF plasmas would be absent if probe protective shields were covered externally by dielectric layers.

Acknowledgments

The authors wish to express their gratitude to Prof. V. Godyak and his colleague B. Alexandrovich for their friendship and effective help in the preparation and arrangement of the present experiment.

This work was supported by the Ministry of Education and Science of the Russian Federation, Project No. 9.9055.2017/BP.

References

[1] Rousseau A, Teboul E, Lang N, Hannemann M, and Ropcke J 2002 *J. Appl. Phys.*, **92**/7, 3463.
[2] Masherov P E, Riaby V A and Godyak V A 2016 *Rev. Sci. Instrum.*, **87,** 02B926.
[3] Masherov P, Riaby V and Abgaryan V 2017 *Plasma Sources Sci. Technol.*, **26,** 015004.
[4] See www.plasmasensors.com. for VGPS Probe System
[5] Godyak V A and Demidov V I 2011 *J. Phys. D: Appl. Phys.*, **44,** 233001.
[6] Riaby V A, Obukhov V A , and Masherov P E 2012 *High Voltage Engineering Suppl.*, **38,** 790 (Proc. 19th Int. Conf. on Gas Discharges and their Applications. Beijing, China).
[7] Masherov P E 2016, Herald of Moscow Aviation Institute, **23**/2, 42.
[8] Riaby V A and Masherov P E 2016, *Proc. of the Rus. Acad. of Sciencies.* Power Engineering, 2, 46.
[9] Vorobiov V V, Kondakov M I, Maliarov A V, Martynova T S, Riaby V A, Savinov V P, Sporykhin A A, and Yakunin V G 1986, Invention Certificate 1429842, Int. Cl. H01L 21/306.
[10] Boyarshinov Yu A, Ermolova N A, Maliarov A V, Riaby V A, Savinov V P, Sporykhin A A, and

Yakunin V G 1988, Invention Certificate 1609392, Int. C l. H01L 21/302.
[11] Iermolova N A, Maliarov A V, Riaby V A, Savinov V P, Sporykhin A A, Sheiko L N, and Yakunin V G 1991, Contr. Papers of the XX ICPIG (Il Ciocco, Italy) Pisa: Felici Editore, **1,** 325.
[12] Kireev V Yu, Kovalevsky V L, Riaby V A, Savinov V P, Sporykhin A A, and Yakunin V G 1992 Rus. Patent RU2024990, Int. C1. 5H01L 21/302.
[13] Alexandrov A F, Riaby V A, Savinov V P, and Yakunin V G 1995 Contr. Papers XI Int. Conf. on Gas Discharges and their Applications (Tokyo, Japan) Tokyo: Part I, 498.
[14] Riaby V A, Savinov V P, Sporykhin A A, and Yakunin V G, 1995 FTIAS Labours, 10, 150.
[15] Riaby V A, Savinov V P, Sporykhin A A, and Yakunin V G 1996 *Microelectronics*, **25/**2, 127.
[16] Riaby V A, Savinov V P, Sporykhin A A, and Yakunin V G 1996 *Microelectronics*, **25/**2, 45.
[17] Alexandrov A F, Riaby V A, Savinov V P, and Sporykhin A A 2005 *Microelectronics*, **34/**1, 21.
[18] Lee H-J, Plaksin V Yu, and Riaby V A 2006 *Proc. Int. Conf. on Advanced Technologies* (Cheju, Rep. of Korea) Cheju: Cheju Nat. Univ., 11.
[19] Bulaeva M N, Kirpichnikov A P, Kravchenko I V, Loeb H V, Masherov P E, Riaby V A, and Tkachenko D P 2012, Herald of Kazan Technological University, **15/**18, 69.
[20] Masherov P E, Obukhov V A, Riaby V A, and Savinov V P 2013 Proc. 21 Intern. Symp. on Plasma Chemistry (Cairns, Australia) Cairns: Austr. Nat. Univ. report 410 [www.ispc-conference.org/ispcproc/ispc21/ID410.pdf].
[21] Godyak V A 2011 *Plasma Sources Sci. Technol.*, **20,** 025004.

Some experiments with the tunnel probe in a low temperature magnetized plasma

J. Kovačič[1], T. Gyergyek[2,1], B. Kavaš[3], M. Vodnik[3], J. Kavčič[2] and J. P. Gunn[4]

[1]Jožef Stefan Institute, SI-1000 Ljubljana, Slovenia
[2]University of Ljubljana, Faculty of Electrical Engineering, SI-1000 Ljubljana, Slovenia
[3]University of Ljubljana, Faculty of Mathematics and Physics, SI-1000, Ljubljana, Slovenia
[4]CEA, IRFM, F-13108 Saint-Paul-Lez-Durance, France

E-mail: jernej.kovacic@ijs.si

Abstract. Experiments were performed using a Tunnel Probe (TP) inside the weakly-ionised plasma of the Linear Magnetized Plasma Device (LMPD). The TP is designed as a concave probe, which should annihilate the problem of sheath expansion in the ion branch of the I-V characteristic. As the ion saturation current is consequently well defined, the ion parallel current and plasma density can be more accurately calculated. Furthermore the ratio between the ion saturation currents on the two collectors (tunnel ring and the back-plate) can be used to derive the electron temperature. The TP has been repeatedly used with success on the former Castor and Tore-Supra tokamaks and will be used on the upgraded version of Tore-supra, namely the WEST tokamak, as well [1, 2]. It was however never used successfully in a low-temperature plasma.

We studied the feasibility of the TP use in a low-temperature plasma for direct measurements of plasma temperature and density. The various probe characteristic dimensions, such as the distance between the two collectors, the aperture size and the probe radius were varied to see influence of the individual probe feature. We also varied the level of magnetization of the charged particle species, the background gas pressure (which influences the electron energy distribution function), the plasma density (important for the ratio between the λ_D and the ion Larmor radius). The sensitivity of the probe alignment to the magnetic field lines was also studied. We found, that the ion saturation current does not necessarily saturate and that the probe works according to expectations only in a limited amount of regimes.

1. Introduction

Electric probes maintain a secure position among the diagnostic tools in plasma environments, be it in technological plasmas, fundamental research or nuclear fusion. They have been used with success from the starts of laboratory plasma physics, since they can provide good temporal and spatial resolution for measuring plasma parameters. Many different and advanced probes have been designed and put to use since the simplest of them, the Langmuir probe (LP), a small cylindrical piece of wire, was inserted into plasma for te first time. However, all probes inherently have some issues, as it is impossible to fulfil all requests regarding the probe size and shape at the same time.

One of the most persisting issues with the convex probe types (like the LP) is the non-saturation of the ion branch of the current-voltage characteristic (I-V). That is because the sheath edge expandes away from the probe surface with increased bias, effectively increasing the collection area for the ions. The increase of the sheath thickness is often in the same order of magnitude as the probe itself, which is due to the fact, that the probes are kept small to reduce its influence on the surrounding plasma. Even for larger, dome shaped probes active in tokamak divertors it was shown that the ion branch often does not saturate [1]. Since the ion saturation current is usually used in calculations of plasma density, Mach number and in measuring of the fluctuations in the tokamak scrape-off-layer, it is unwise to neglect the problem of the sheath expansion.

A new probe, a so-called tunnel probe, appeared in use on tokamaks a some years ago [2] and it manages to overcome the difficulties arising with the sheath expansion by strongly defining the collection area. The probe is of a rather simple concave design, reminding us of an inverted Langmuir probe.

Our goal was to construct a tunnel probe, that we would be able to use inside our Linear Magnetized Plasma Device (LMPD), located at Jožef Stefan Institute, Ljubljana.

2. Tunnel probe

Therefore, the tunnel probe was invented consisting of two collectors, a ring (tunnel collector) followed by the backplate. The collectors are electrically isolated from each other. The orifice of the probe should always be normal to the magnetic field in order to function properly. In that way, the surface of the tunnel shall always be parallel to the magnetic field and the backplate shall always be normal to the magnetic field.

The idea behind this design is that the probe's collectors enclosed in a cylindrical cavity can only be reached by charged plasma particles through that single orifice on the front of the probe. The expansion of the sheath (and pre-sheaths) with biasing of the collectors should therefore appear only inside the cavity.

The standard mode of operation of the TP is having both collectors biased highly negatively in order to repel all the electrons and collect all the ions coming through the orifice. There are two well defined parameters that you can obtain from a single measurement. The first one is obviously the parallel ion current density, where we can generally say that

$$j_{\parallel,+} = \frac{I_{probe}}{A_c} = Ze_0 n_k v_k \tag{1}$$

where Z is the ion charge number, e_0 is the elementary charge, n_k and v_k are ion density and velocity at a designated point in front of the collector, where we are able to employ the conservation of the ion flux. In the most common approach we use the Debye sheath edge as the point, where we assume that the ion velocity equals the ion acoustic speed and the density evaluation is based on the Boltzmann factor. It is obvious, that even such a simplistic approach has an issue with defining the collection area A_c, when we move from a 1-d case to a realistic cylindrical case with the Langmuir probe, where the expansion of the sheath is present. The Child-Langmuir law can be used to describe the enlargement of the sheath due to the voltage between the bulk plasma and the collector. In the simplest form it relates the maximum current between planar electrodes and the voltage between them.

$$j_{CL} = \frac{4\epsilon_0}{9}\sqrt{2e_i/m_i}\frac{V_b^{3/2}}{d^2} \tag{2}$$

The second quantity, that we can extract from such measurement is the temperature of the electrons, which seems strange at first, since no electrons are collected at such negative bias. However, electrons do enter the probe cavity and the processes of plasma-wall transition are contained inside it. The ratio of the collection of the ions is defined by two competing mechanisms, namely the magnetization of the ions and the electric force that drags the ions to the collectors. By increasing the magnetization we increase the chance that the ion entering the cavity will reach the back plate. On the other hand, if we increase the sheath thickness in front of the tunnel (with bias or by changing plasma properties), more ions will be deflected from their orbits into the ring collector.

3. Experimental setup

The experiments were conducted in the Linear Magnetized Plasma Device (LMPD), located at the Jožef Stefan Institute, Ljubljana, Slovenia. The plasma in LMPD is a hot filament discharge. The plasma chamber is a cylindrical tube with length $l_{LPMD} = 1.5m$ and diameter $d_{LMPD} = 17cm$. The plasma is confined by an axial magnetic field of up to $B = 0.4$ T using solenoid with 14 coils.

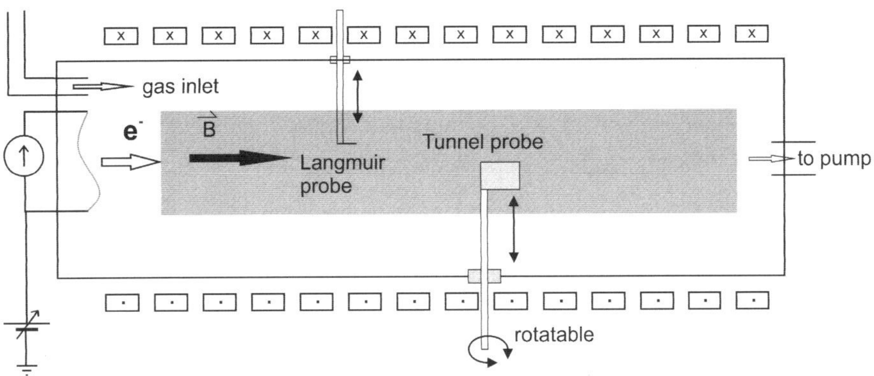

Figure 1. Schematic of the Linear Magnetized Plasma Device - LMPD.

The LMPD is a flexible device, with 9 radial ports and 1 axial port. Different working gases (H_2, He, Ar, O_2, N_2) can be used and the working gas pressure can vary from around p= 10^{-4}mbar to around p= 10^{-2}mbar. Discharge voltage U_d and dicharge current I_d can also be varied. Using these parameters we can alter the electron energy distribution function (EEDF) and the density of plasma, which both, in turn, change the sheath properties inside the tunnel and affect the ratio of ion currents on the two collectors.

As we mentioned before, our main goal was to explore the possibility of use of the tunnel probe inside a low temperature plasma device for measuring the *true* ion saturation current density and the electron temperature. In order to do that, two tasks were in front of us: *i. proper saturation of the ion branch of the I-V* had to be reached, which is not a given in case of convex Langmuir probes and *ii. distinction in the ratio of currents collected by the two collectors* had to be reached for different electron temperatures, which would mean, that we can separate different plasmas.

Since the TP method is applied using a simple model with a firm set of limitations, it is the geometry of the probe that has to be morphed accordingly to fit the probe into the model. The original author of the probe made several hundreds of PIC simulations for various plasma parameters for the scrape-off layer (SOL) plasma of the Tore-Supra and CASTOR tokamaks and chose the most suitable dimensions of the probe. However, the typical SOL densities are about two orders of magnitude above the ones in LMPD, as are the ion temperatures and the magnetic field density.

Nevertheless, we were able to find a suitable operational regime in LMPD. Namely, one of the more important parameters, the degree of magnetization of the ions (ion Larmor radius $r_{L,i}$ and the ion cyclotron pitch $l_{c,i}$), can be reached due to the low temperature of the ions in the

LMPD (we estimate it to be around $T_i = 0.05$ eV). Because of that we can reach the degree of magnetization similar to Tore-supra even with 10 times lower magnetic field. For that reason, we also chose He instead of Ar ions, since lower mass decreases the $r_{L,i}$. If magnetization of the ions is too low, none will reach the backplate.

Another important parameter is the Debye length of the plasma ($\lambda_D = \sqrt{\frac{\epsilon_0 k T_e}{n_e e_0^2}}$). Here again the low density of the LMPD plasma comparing to Tore-supra SOL plasma is significantly compensated by the lower temperature of its electrons. It is important, that the Debye length is significantly smaller than the diameter of the probe tunnel $\lambda_D \ll 2r_{TP}$, otherwise the sheath will expand throughout the cavity and the ions will not reach the backplate. Furthermore, with larger Debye lengths, the sheath can nonetheless start expanding outside of the probe's cavity.

In the Table 1, we have the comparison of relevant parameters.

Table 1. Comparison of relevant parameters of Tore-supra SOL plasma and LMPD He plasma.

	Tore-supra	LMPD (He)
B	3.5 T	0.4 T
T_e	20 eV	2 eV
$r_{L,i}$	$2.5 \cdot 10^{-4}$	$2 \cdot 10^{-4} m$
$l_{c,i}$	$1.6 \cdot 10^{-3} m$	$1 \cdot 10^{-3} m$
n	$5 \cdot 10^{18}\ m^{-3}$	$5 \cdot 10^{16}\ m^{-3}$
λ_D	$1.5 \cdot 10^{-5} m$	$5 \cdot 10^{-5} m$

Consequently we found the optimal dimensions for our configuration and they are written in Table 2. This dimensions are also presented in fig. 2.

Table 2. Dimenisons of the tunnel probe used in the LMPD.

l_{tunnel}	r_{tunnel}	$r_{backplate}$	$r_{shutter}$	l_{spacer}
2 mm	3 mm	3 mm	2 mm	0.4 mm

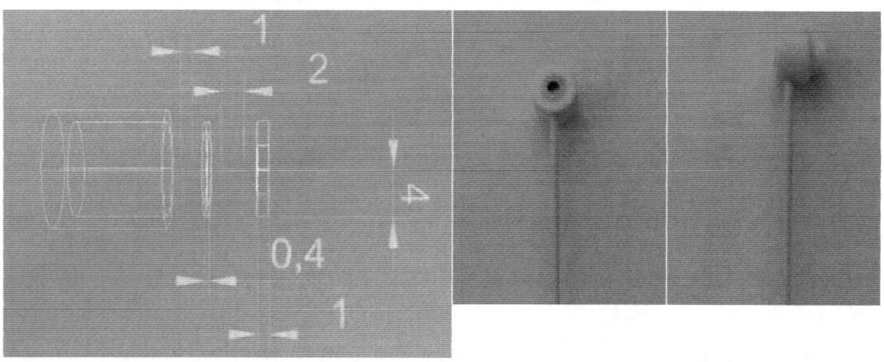

Figure 2. Probe dimensions (left), pictures of the used probe (right).

Collectors in the probe were made of copper, the casing and the shutter were made of PTFE and the probe holder was made of Al_2O_3.

As mentioned before, the TP used in CASTOR and Tore-supra tokamaks was calibrated using data derived from PIC simulation. Since such data was not available for our configuration of probe and plasma, we had to "calibrate" the measurements to a different independent source for temperature measurements. We chose a Langmuir probe, which was installed in a different radial port inside LMPD at the same time, so we were able to do TP and LP measurements in practically the same plasma.

We altered the temperature of the electrons, or better - the EEDF, by changing the gas pressure. At higher pressures the EEDF is Maxwellian, since all the primary electrons suffer many collisions and get thermalized. By decreasing the gas pressure the temperature of electrons starts rising and then splits into two groups - essentially constructing a bi-Maxwellian distribution function. We have the thermalized group at around $T_{e,l} = 2\text{eV}$ and the fast group of electrons at around $T_{e,h} = 15 - 20\text{eV}$.

So, which temperature are we then measuring with the TP? The deflection point for the ions, which defines the collecting area, is definitely somewhere close to the sheath entrance, which means, that according to theory [5], we should consider the effective screening temperature T_e^*. The T_e^* is the temperature which fulfils the Bohm criterion. It is defined by the following equation:

$$\frac{1}{n_j}\frac{dn_j}{d_\Phi} = -\frac{e_0}{kT_e^*} \tag{3}$$

Here, the n_j is the sum of densities of all negative charge species. We made some estimates of the densities of the two electron groups using the first derivative probe method from LP [7]. Unfortunately the LP was oriented perpendicular to the magnetic field, therefore we had the magnetization parameter only around $\Psi_0 \sim 40$. From the measurements we saw, that the density of the fast electrons never exceeded 15% of the thermal population. According to [6] we could now neglect the fast group in the effective temperature calculations, meaning that our $T_e^* = T_{e,l}$.

4. Results and discussion

4.1. Susceptibilty to the probe orientation

First thing of interest was the susceptibility of the measurements to the probe orientation with respect to the magnetic field. The probe should be oriented exactly along the magnetic field axis in order to work properly. We made a sweep from $+90^o$ to -90° with 0° being along the magnetic field. We can see the ion saturation currents to the backplate and the tunnel on the left graph in fig. 3. Here we can see a slight offset of the 0° position, which we corrected afterwards

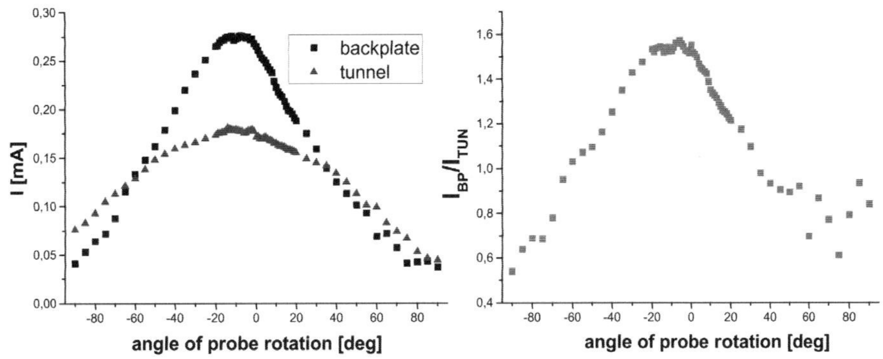

Figure 3. Helium discharge, $p = 4 \cdot 10^{-4}$, $I_d = 2.2A$, $U_d = 60V$, $B = 400mT$, I_{sat}^-.

IOP Conf. Series: Journal of Physics: Conf. Series **958** (2018) 012007 doi:10.1088/1742-6596/958/1/012007

for subsequent measurements. Since the interesting quantity is in fact the ratio of both currents, we had a look at how it varies with the tilting angle (right graph in fig. 3)

Here we can see, that the ratio is not significantly affected around the middle position, giving us an angle of confidence of at least $\pm 5^o$, which is quite good.

4.2. Measuring the ion parallel saturation current density $J_{\|}$ with tunnel probe

Next we shall present different cases of configurations of the TP and the plasma parameters in Fig. 4 to present the issues associated with the dimensioning of the probe. In the leftmost picture we can see, that the ion branch of the I-V does not saturate. Furthermore, we can notice that the current to the tunnel is significantly higher than the current to the backplate. This means that the probe diameter (same probe dimensions as presented in Table 2) is to small. However, significantly increasing the probe size would be contradictory to the rule of avoiding plasma perturbations by the probe. Such probe can therefore only work with higher magnetization of the ions. In the middle picture of fig. 4 we again do not have saturation of the ion branch. In this case the probe had a radius of only $r_{tunnel} = 1$ mm. In this case we can see, that almost all ions are collected by the tunnel. We also tried the same probe with a longer tunnel, however it was very difficult to set the correct length where the current to the tunnel collector would be of a comparable size to the tunnel to the backplate.

We got best results when we used the probe with dimensions stated in Table 2. At high enough magnetic field densities we would then obtain I-V's similar to those presented in [4] as can be seen in the rightmost picure of fig. 4. We would now be able to correctly measure the parallel ion saturation current density.

4.3. Measuring the electron temperature T_e with tunnel probe

Next up were the electron temperature measurements. Here we would proceed with simultaneous measurements of I-V's with the Langmuir probe and collector current ratio with the tunnel probe. We would vary the temperature by changing the gas pressure. The temperature would be obtained from the slope of the I-V characteristic. As discussed before, with lower gas pressure the temperature of the thermalized electrons would rise (as well as the effective screening electron temperature).

As we see in the first graph in fig. 5, there is a logarithmic relation between the electron temperature and the working gas pressure. Similarly, there is a logarithmic relation also between the gas pressure and the ratio of the currents to the tunnel collector I_{TUN} and to the backplate collector I_{BP}, respectively (middle graph in fig. 5). If we join these two graphs, we obtain the relationship between the electron temperature T_e and the current ratio I_{BP}/I_{TUN}. The resulting dependence could be described as linear, which means, that in principal it is possible to use the

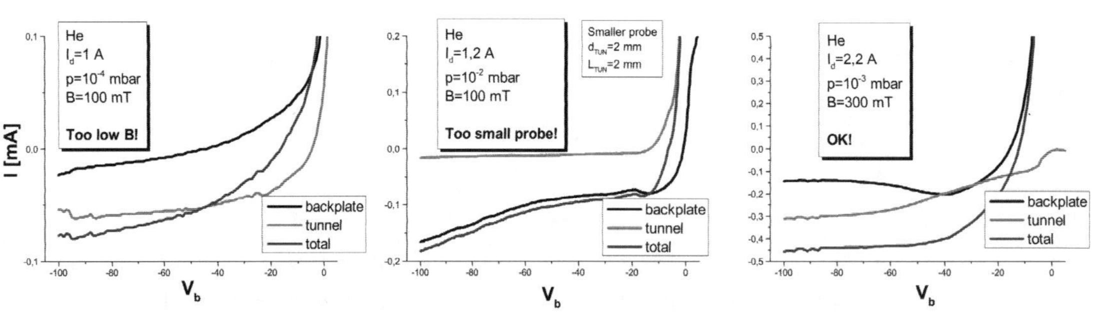

Figure 4. Tunnel probe - 2 cases of wrong probe dimensioning (left and middle) and a case of finding the correct dimensions (right).

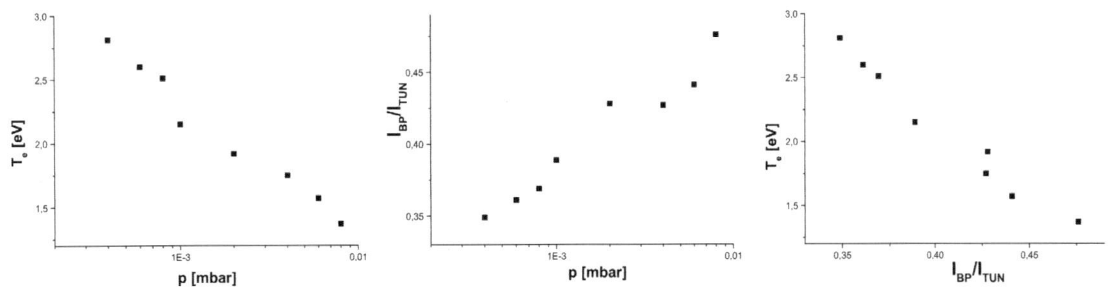

Figure 5. Measuring the temperature T_e using the tunnel probe. The T_e was varied by changing the gas pressure in the vessel. Helium discharge, $p = 4 \cdot 10^{-4}$ - $8 \cdot 10^{-3}$, $I_d = 2.2A$, $U_d = 60V$, $B = 400mT$.

TP for direct measurement of electron temperature, as long as it has been calibrated to another method for electron temperature evaluation.

The results are as expected. For higher electron temperatures the Debye shielding is less effective. Because of that the sheaths around the tunnel collector expand more towards the axis of the tunnel and I_{TUN} rises in comparison to I_{BP}. The TP appears to work like that also in the tokamaks. We realize that this is only one of the mechanisms, since collisions, return current heating etc. should be taken into account, but it matches the results nonetheless.

4.4. Saturation of the electron branch of the I-V characteristic
Lastly we have had a look into the electron branch of th I-V characteristic to see if electron saturation is possible for this type of probe. This is something, which was not successfully done in the tokamak TP. In fig. 6 we see that electron branch does indeed saturate for high enough bias and as expected, the backplate collects almost all of the current, since the electron cyclotron radius is very small comparing to the probe radius. This opens new possibilities as electron saturation current could be used for calculations as well.

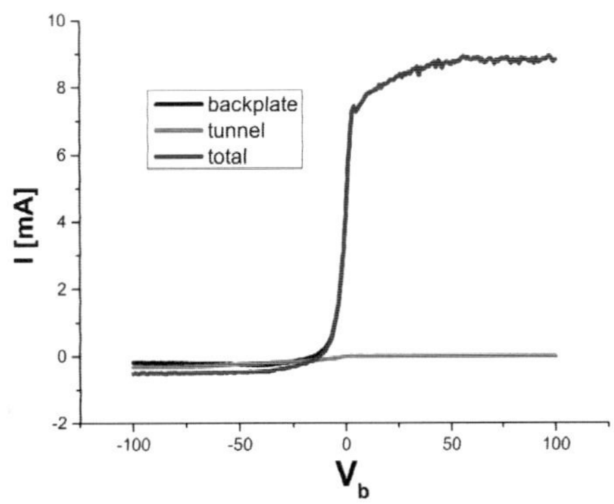

Figure 6. The full current-voltage characteristic with tunnel probe. Saturation of currents is seen in the electron and the ion branch. Helium discharge, $p = 1 \cdot 10^{-3}$, $I_d = 2A$, $U_d = 60V$, $B = 300mT$.

5. References
[1] Dejarnac R, Gunn J P, Stöckel J, Adamek J, Brotánkova J and Ioniţă C 2000 *Plasma Phys. Control. Fusion* **49** 1791

[2] Gunn J P, Devyncek P, Pascal J-Y, Adámek J, Ďuran I, Hron I, Stöckel J, Žaček F, Bařina O, Hrach R, Vicher M and van Oost G, 2002 *Czech. J. Physics* **52** 1107
[3] Gunn J P, Schrittwieser R, Balan P, Ioniţă C, Stöckel J, Adámek J, Ďuran I, Hron M, Pánek R, Bařina O, Hrach R, Vicher M, van OoSt G, Van Rompuy T and Martines E 2004 *Rev. Sci. Instrum.* **75** 4328
[4] Gunn J P, Dejarnac R, Stöckel J 2016 *J. Phys.: Conf. Series* **700** 012018
[5] Riemann K-U 2000 *J. Tech. Phys.* **41** 89
[6] Čerček M, Gyergyek T, Stanojević M 1999 *Contrib. Plasma Phys.* **39** 541
[7] Popov Tsv, Ivanova P, Dimitrova M, Kovačič J, Gyergyek T, Čerček M, 2012 *Plasma Sources Sci. Technol.* **21** 025004